Geospatial Services and Applications for the Internet

Geospatial Services and Applications for the Internet

Edited by

John T. Sample
Kevin Shaw
Naval Research Laboratory
Stennis Space Center, MS, USA

Shengru Tu
Mahdi Abdelguerfi
Dept. of Computer Science
University of New Orleans, LA, USA

Editors
John T. Sample
Naval Research Laboratory
1005 Balch Blvd.
Stennis Space Center, MS 39529
John.sample@nrlssc.navy.mil

Kevin Shaw
Naval Research Laboratory
1005 Balch Blvd.
Stennis Space Center, MS 39529
shaw@nrlssc.navy.mil

Shengru Tu
Dept. of Computer Science
2000 Lakeshore Drive
New Orleans, LA 70148
shengru@cs.uno.edu

Mahdi Abdelguerfi
Dept. of Computer Science
2000 Lakeshore Drive
New Orleans, LA 70148
mahdi@cs.uno.edu

ISBN-13: 978-1-4419-4510-5 e-ISBN-13: 978-0-387-74674-6

Printed on acid-free paper

9 8 7 6 5 4 3 2 1

springer.com

Contents

List of Figures

List of Tables

Preface

The use of geospatial technologies has become ubiquitous since the leading Internet vendors delivered a number of popular map Web sites. Today, businesses are either migrating location-aware capabilities into their information systems, or expanding existing Geospatial Information Systems (GIS) into enterprise-wide solutions. For example, GIS has been an essential component of the workflow and decision support system of some Homeland Security programs. GIS experts have been pursuing a number of vastly large pilot geospatial systems such as the Global Earth Observation System of Systems (GEOSS), and the Federated Earth Observation Missions (FedEO) Pilot programs.

As enterprise information systems are evolving toward service-oriented architecture (SOA), geospatial technologies have also been evolving along the same direction. The SOA has been best fulfilled by the Web service technology for general purposes, and the OGC Web Services Common (OWS) Specification for geospatial information systems. The service-oriented approach enhances GIS with programmable, automatic interactions between the data resources and the client applications across the Internet. Consequently, a new category of services can be added to GIS: business process services including work-flow realization and location-aware decision-making assistance. In addition to the three traditional categories of GIS services namely the data services, data transformation services, and catalog/registry services, the business process service is a significant new category.

In this book, we have gathered a coherent collection of techniques and methodologies for the development of geospatial information systems in the Internet era. This manuscript is to appeal to the researchers and practitioners who are either geospatial information service providers or developers of the client components that utilize geospatial information. The nine chapters of this manuscript are divided into two parts which address the challenges coming from two key areas, namely service-oriented architectural design and implementation, and intelligent spatial query processing.

Part 1 – System Design and Implementation

A central concern of any geospatial services over the Internet is scalability and the user-perspective of performance, which requires an integrated design at each of the levels in the data delivery across the servers and the clients. In Chapter 1, Brabec and Samet's hierarchical infrastructure for Internet mapping services pays a special attention to the approaches based on transmission of vector data. Compared to the current popular image-only (raster-based) approach, their vector-based approach can significantly enrich the basis of a foreseeable wave of intelligent geospatial applications because vector data would make it possible for client-side widgets to support intelligent manipulation of spatial data. This is a systematic study of architectures for delivering map data including the "direct server access" and "adding auxiliary servers" as well as a hybrid approach. Each of them is further broken down into sub-categories depending on the use of memory and caching techniques. Every design choice is justified by experiments using real world data; the experiments are clearly explained with adequate details.

In Chapter 2, Diaz, Granell and Gould present a case study of the design and implementation of a concrete Web-based geospatial information system for hydrological applications. This is an example illustrating what is possible with some of the Web service technologies. For example, the system allows the hydrologists to interact directly with the underlying hydrological model; it allows the expert users to load their own datasets for an area of interest. The latest OGC Web Processing Service (WPS) specification is applied to the design and implementation of the system, which experimented the three-method approach (getCapabilites, describeProcess, and execute methods) in a customized WPS client-side API along with Java servlets and JSPs. An interesting feature of this chapter is the use of Google Maps API in the client-side applications. It is invaluable for this case study to point out that, after numerous standards have been established, it is still difficult to achieve interoperability by just relying on the existing standards because vendor implementations of a standard often differ from one another.

Realizing the complexity and difficulties in building Web-based GIS applications, Milosavljevic, Dordevic-Kajan and Stoimenov have developed, in Chapter 3, an application framework called "GinisWeb." The goal of this framework is to assist in the rapid development of Web GIS applications and to facilitate the access and sharing of geospatial data through OGC Web standards such as WMS and WFS. A difference between the GinisWeb applications and the purely standard-based GIS applications, such as the hydrological applications presented by Diaz's team in Chapter 2, lies in the approach to encoding of application descriptions including the metadata of geospatial datasets, data organization, data layers and data services (acquiring, storing and querying). The GinisWeb framework uses a "homemade" XML language – Ginis Application Definition Language (GADL). Such a technical choice led to a more concise application description and easier deployment because the basic data unit in GADL is layer. In doing so, the GinisWeb framework shifts away from the OGC WPS standard. However, even for the reader who may prefer to follow the standards strictly, this chapter is a worthy reading because it presents a series of technical choices in system design (with a rich set of UML diagrams) that are typical and critical in GIS design.

Interoperability has been the primary priority of OGC's efforts. The OGC Interoperability Program has been advocating integration between OGC services and W3C SOAP-based services for years. The simple conversion method is an early approach in which each of the required operations (such as `GetCapabilities` and `GetMap`) of a WMS service was directly wrapped in a SOAP-based service.

The connectivity issue was addressed; any SOAP-based service client program can call the wrapped services. However, the client program will not be able to use the map service unless it is as capable as a WMS client because the actual information about the map layers have to be retrieved through the `GetCapabilities` operation defined in WMS. In Chapter 4, the approach to the automation of mapping OGC services to SOAP-based services exposes all the map layers of the WMS service in multiple searchable service descriptions – WSDL, and achieves the full-scale access from the W3C service to the original OGC services. Not only can ordinary service client programs (without WMS capability) retrieve any layer provided by the WMS service, but the client programs will also be able to assemble any combination of layers from multiple WMS services.

In Chapter 5, Pleasea presents a field report of one of such systems – the OnEarth WMS server at the Jet Propulsion Laboratory (JPL). This system is designed to provide fast access via WMS and the Keyhole Markup Language (KML) to very large sets of satellite imagery. It utilizes a cluster of systems to break the processing work up. The author describes the Web and image processing pipeline and provides a detailed description of the software and hardware configuration behind this fast and highly available system. The current status and development history are described. A number of technical choices made over the eight years of the system history can serve as lessons learned from the large-scale geospatial system.

Part 2 – Intelligent Spatial Query Processing

As enterprise information systems are integrated into regional applications, geospatial information naturally takes the role as the convergence platform. Early efforts in fulfilling data integration and interoperability often paid more attention to syntactic heterogeneity and connectivity issues such as message formats. More recent research and development efforts start addressing more difficult problems such as semantic heterogeneity. In Chapter 6, Cruz and Xiao present a systematic approach to geospatial query processing over semantically heterogeneous data sources based on a set of ontology alignment algorithms. Their approach was for a specific application – the land usage system, in which the data resources collected by different counties (in Wisconsin) and agents differ from each other in many ways. Cruz and Xiao carried out instance-level mappings over the taxonomy-like ontologies about land usage consisting of the *subClassOf* relationship only, and the mappings over the schema-like ontologies which are associated with the structures of the local resources. Driven by the complexity existing in the real-world data resources, they developed a set of automatic query rewriting methods to support query processing that are involved in distributed, heterogeneous sources in either the global-to-local or local-to-global mode.

Concentrating on a very specific problem – mapping vernacular terms to location coordinates, Adeva presents a statistical text mining technique in Chapter 7. Being independent of languages and grammars is an advantage of this technique, which can also map the location terms that are often-used but not included in vernacular collections. This technique depends on a prior-built knowledge set which is the coordinates and the descriptions of all the locations within a geographical area, in which the words are weighted using statistics. The information in the knowledge set is based on external information such as gazetteer, encyclopedias or information from the Web. The knowledge set is transformed into a reduced, traditional vector space model. For a static vernacular term set, the knowledge needs to be built once only. After it is built, the mapping process is a computationally light operation. A weak spot of this technique lies in the mapping accuracy which is subject to the granularity of the locations provided by the gazetteer.

A personalized processing can improve the suitability of query answers. This can be considered a "stretching" extension of query processing. In Chapter 8, Aoidh, Bertolotto and Wilson present a methodology for personalizing location-aware services based on the implicit measures of user interactions with digital maps – the mouse movements. The effectiveness of this approach is measured using three metrics, namely the rank accuracy (the accuracy of the ranks assigned to the objects by the algorithm compared to the ranks given by the user in survey), the absolute preference (an ordered rank list showing the level of interaction associated with every objects of interest in the 0 to 100 scale), and the relative preference (the ratio of the score difference between two adjacent objects in the computed rank list). With the dataset used in the evaluation, the experimental results show that the mouse movement analysis over spatial data is a technique for accurately inferring users' interests. This is an innovative approach that could prepare for a better display of spatial query for individual users. This chapter exemplifies an effort to attract more attention from the human-computer interaction community toward GIS applications such as digital maps.

We called upon many experts in the fields to assist with the reviewing process of the submitted chapters. We would like to extend our thanks to them for their rigorous reviews which improved many chapters in this manuscript. Special thanks to Fangfang Liu for her assistance with the formatting of this manuscript.

Chapter 1: Hierarchical Infrastructure for Internet Mapping Services

Frantisek Brabec and Hanan Samet*

Department of Computer Science
University of Maryland
College Park, Maryland 20742, USA
brabec@cs.umd.edu and hjs@cs.umd.edu

Abstract For years, the access to Internet-based public mapping services provided by vendors such as MapQuest or MapsOnUs has changed little. The mapping service would generate maps of the viewed areas in raster format and transfer them in the form of images embedded in Web pages to remote users. This approach is suboptimal for users who plan to explore a given area in more detail as the same data may be sent to the users repeatedly. In mid 2005, Google Maps and MS Virtual Earth improved upon this approach by dividing the images into smaller tiles which allows many of them to be reused in subsequent panning. This increases performance of such mapping systems substantially. In both cases, however, the client only has access to data converted in its raster format which prevents it from querying or re-processing the data locally. We investigate this opportunity for further improvement in providing the client with map data in vector format so that it can perform some operations locally without accessing the server. We focus on finding strategies for distributing of work between the server, clients, and possibly other entities introduced into the model for query evaluation and data management. We address issues of scalability for clients that have only limited access to system resources (e.g., a Java applet). We compare performance of the vector-based system with raster-based systems, both traditional (e.g., MapQuest) and tiled methods (e.g., Google Maps) for a set of common basic operations consisting of fine and fast scrolling and zooming (both in and out).

---

* The support of the National Science Foundation under Grants EIA-00-91474, CCF-0515241, and IIS-0713501; Microsoft Research; and the University of Maryland General Research Board is gratefully acknowledged.

1 Introduction

Technological advances in recent years have opened ways for easier creation of spatial data. Vast amounts of data are collected daily by both governmental institutions (e.g., USGS, NASA) and commercial entities (e.g., IKONOS) for a wide range of scientific applications (e.g., [18]).

The motivation is the increased popularity and affordability across the spectrum of collection methods, ranging from personal GPS units to satellite systems. Many collection methods such as satellite systems produce data in raster format. Often, such raster data is analyzed by researchers directly, while at other times such data is used to produce a final dataset in vector format. With rapidly increasing data supplies, more applications for the data are being developed that interest a wider consumer base. The increasing popularity of spatial data viewers and query tools with end users introduces a requirement for methods to allow these users to access this data for viewing and querying instantly and without much effort. Our work focuses on providing remote access to vector-based spatial data, rather than raster data.

Traditionally, common spatial databases and Geographic Information Systems (GIS) such as ESRI's ArcInfo are designed to be stand-alone products. The spatial database is kept on the same computer or local area network from which it is visualized and queried. There are, however, many applications where a more distributed approach is desirable. In these cases, the database is maintained in one location, while users need to work with it from possibly distant places over the network (e.g., the public Internet). A common approach for providing access to remote spatial databases adopted by numerous Web-based mapping service vendors (e.g., MapQuest [8]) performs all the operations on the server side, and then transfers only bitmaps that represent results of user queries and commands. Although this solution only requires minimal hardware and software resources on the client site, the resulting product is severely limited in the available functionality and response time (each user action results in a new bitmap being transferred to the client). Naturally, the drawbacks of this traditional approach have been identified and work has started to improve the performance of remote spatial access using both raster [1] and vector [19] approaches. Similar issues were addressed in a component-based Web GIS [12] tool by adding a spatial caching framework.

Providing efficient data flow between a given spatial server and individual clients is not the only problem that needs addressing. In many scenarios, data originates from multiple providers and the information they offer needs to be aggregated before being presented to the end user. Similarly, multiple spatial servers may be involved for redundancy or load balancing. Such topics have been explored in other work, when the providers' hosting environment remains stable [20] and for more dynamic peer-to-peer arrangements [11].

In our research, we explore new ways of allowing visualization of both spatial and nonspatial data stored in a central server database on a simple client connected to this server by a possibly slow and unreliable connection. We develop a new vector-based client-server approach as a response to some of these drawbacks of traditional solutions. Our system aims to partition the workload between the client and the server in such a manner that the user's experience with the system is interactive, with minimal delay between the user action and appropriate response. We consider scenarios where bringing in auxiliary servers would improve the performance of the system. The design works around potential bottlenecks for the information transfer such as the limited network bandwidth or resources available on the client computer. To support multiple concurrent clients, limited resources on the server must also be considered. We will see that the performance of our vector approach is comparable and at times better than the latest raster-based methods.

The rest of the chapter is organized as follows. Section 2 reviews existing commonly used methods for remotely accessing spatial databases. Section 3 discusses our architecture based on pure client-server approach. Given a client that communicates directly to a server, we examine different deployment options and describe several methods that improve the performance that can be achieved in this environment. Section 4 extends the basic client-server approach by adding auxiliary servers. Such servers can be used as temporary data storage between the client and the server. We present typical deployment scenarios when this would be beneficial, as well as present methods for using this arrangement to further speed up its performance. Section 5 combines all the different design options and speed-up methods together, performs evaluations, and discusses how to choose the optimal deployment method for given specific usage scenarios. Section 6 shows results of experiments comparing performance of our method and existing established raster-based remote access methods. Finally, Section 7 draws some conclusions and proposes topics for further research.

2 Internet Mapping Services

Many vendors that provide access to maps over the Web utilize an approach where server-generated bitmaps are sent to the Web browser client for viewing. The typical example of providers of such services are vendors such as MapQuest [8] for street maps based on addresses; or TopoZone [7] for topographical maps. Their approach is simple, the server receives a location description (e.g., a street address, name of a place, etc), it queries its spatial database, retrieves a map, converts it into a bitmap image and sends it back to the user (their browser). The map retrieved from the spatial database may be in vector (MapQuest) or raster (TopoZone) format. In either case, it gets rasterized or subsampled respectively before sending the data over the network to user's browser.

This approach requires very little support from the client site, typically just a Web-browser equipped computer or network appliance. The drawback of this solution is that it quickly reaches its usability limitations when more serious work is attempted. Such poorly supported operations include even basic zooming in or out or panning not to mention running queries. In particular, actions such as zooming or panning are very cumbersome with performance bordering unacceptable for many users as the response time is determined by the amount of data that needs to be transferred every time a new view is requested. Other operations such as querying the database beyond displaying all objects within a certain rectangle are not supported at all.

An interesting enhanced raster-based design has recently been presented by Google [1] and Microsoft [3]. Similar to MapQuest, Google Maps and Microsoft Virtual Earth (and its predecessor TerraServer [10]) services are raster-based as is also NASA's World Wind [5] which besides working with NASA's own data it also makes use of data from TerraServer. However, these services do not send a single image covering the whole viewable area every time there is a need for an update. Instead, the viewable map is divided into a grid of smaller image cells. When a panning operation is executed, there is no need to download a new image that represents the whole viewable area. Only cells covering the area that just became visible need to be downloaded, others are reused by simply moving them on the screen.

As an alternative to these raster-based systems, we consider the SAND Internet Browser [17] — a Java application that represents the client piece of our vector-based client-server solution for facilitation of remote access to spatial databases.

3 Direct Server Access

Traditionally, a client-server computing paradigm only involves two computers — the client and the server (obviously ignoring computers and devices in between the two that simply route or shape the traffic between them, such as routers, firewalls, etc). We examine such a scenario as well as scenarios that involve other auxiliary servers.

3.1 Pure Client-Server Design

The simplest and most common design for the client-server architecture makes individual tasks such as data management, image rendering, and query evaluation the responsibility of either the client or the server. When the spatial database application is implemented in this manner, the server handles all the data management and query evaluation. The client only facilitates data visualization while maintaining connectivity to the server. In this scenario, the client simply translates user input into queries and transmits them to the server. It can also receive data sent by the server and visualize it. There is no data storage or processing on the client beyond these basic functions. Note that this design corresponds to the way in which many popular Web-based mapping services such as MapQuest operate.

This approach's advantage is that most users can utilize the service with the resources that they already have — that is, a networked machine with a Web browser. Users do not need to install or set up any additional hardware or software. However, this approach's main drawback is that clients need to communicate with the server each time users request even the simplest operation. This can slow down users experience significantly if the network throughput and latency are a limiting factor or if the server is heavily loaded.

3.2 Memory-Based Caching in the Client

The first method that improves upon the basic design is one where the client utilizes some of its own main memory to store (cache) some of the data in the central database. This allows the client in some cases to rely on its own data repository to handle some of the user's requests thus cutting back on the network utilization and improving the system's responsiveness. Naturally, the spatial data stored on the client must be spatially indexed for fast access. Note that in this approach it is no longer possible to use the standard Web browser as a mere image viewer. In particular, custom code needs to be loaded onto the client to facilitate the operations to be performed there. The Java environment has emerged in the past years as a platform of choice for most types of lightweight cross-platform applications. The maximum amount of data to be stored on the client to optimize the overall performance depends on the client's available resources. The rationale for this design is for the client to fetch the requested data via fast memory-only operations whenever possible. This is more efficient than retrieving the same data over the network from the central server.

Operations performed by the mapping system are primarily client-driven, i.e., any operation performed on either the client or the server is in response to some user-generated input. To minimize the amount of data that needs to be transferred from the server to the client in response to each event on the client side, various techniques were developed and implemented in the form of the SAND Internet Browser. To keep the amount of traffic between the client and the server low, we cache some data on the client in case the user requests another operation on data in the same area. We store the data in their original vector format rather than the resulting bitmaps so that the client is able to generate new views and process some types of queries locally without having to request additional data from the server.

3.3 Internal Spatial Data Structures

The spatial data is stored on the client using a PMR quadtree [13] spatial data structure. This structure subdivides the plane into quadrants such that if an object is inserted into a certain quadrant, then if the quadrant already contains more than a predefined threshold of other objects, then the quadrant is split into its four children once and only once and the objects are reinserted into the children. Thus, the objects are always stored in the leaf nodes of this quadtree. We establish and maintain the maximum amount of data that can be cached on the client in order not to overwhelm or crash the client platform. Each leaf node of the PMR quadtree also contains a time stamp indicating when it was accessed (displayed) last. Together with the PMR quadtree containing the spatial data, we also maintain pointers to all of the PMR quadtree leaf nodes using a variant of a binary heap data structure. The key for this tree is the time stamp stored in the PMR quadtree leaf nodes. This is shown in Figure 1.1.

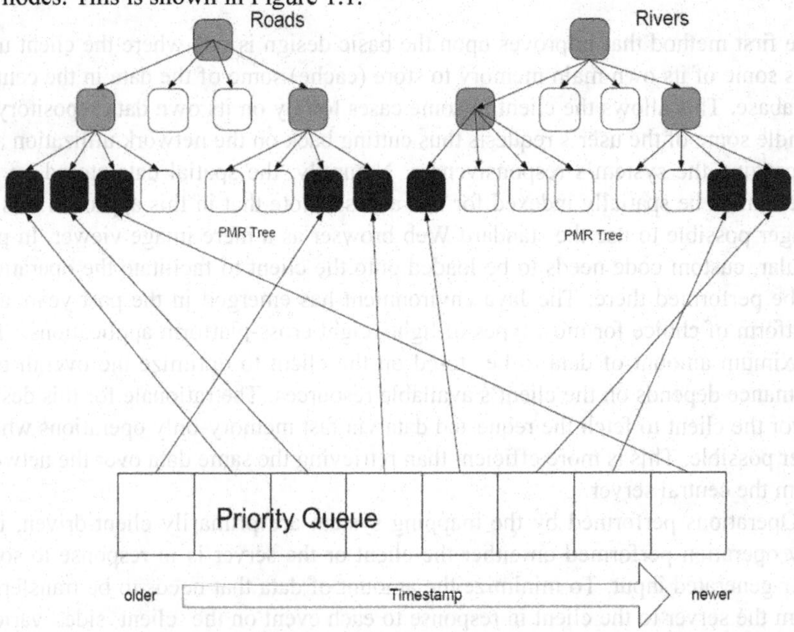

Figure 1. 1 Individual spatial data layers are stored in separate PMR quadtrees. A priority queue shared by all of them maintains ordering of all the PMR leaves for all the PMR quadtrees based on the time of their last viewing.

This structure enables quick insertions, deletions and locating the pointer representing the PMR node with the oldest time stamp. This arrangement facilitates our caching mechanism. When we need to make more memory available for additional data, we use the least-recently-used (LRU) caching mechanism to delete as many PMR leaf nodes linked from the top of the binary heap data structure as necessary. If all four children of some internal PMR quadtree node are removed, then the quadtree automatically collapses and the internal node becomes an empty leaf node. A flag in each node indicates whether the node represents an area that is actually empty (valid node) or whether the node is empty because its elements are not available in the memory (e.g., page fault, invalid node).

Note that using this mechanism, the entire quadrant has to be contained in the memory for its node to be valid. This may be too inefficient as if we continuously work with only part of the quadrant and have no need to load the rest of the quadrant, then the node would never be marked as valid and the data from the part in which we are interested would be reloaded over and over. To prevent this, we add another field in each node indicating which part of it is actually valid. Thus, if we loaded data for only part of the quadrant, we mark the quadrant as valid but indicate which part of it is actually valid (i.e., the intersection of the quadrant and the query window). The next time we need to access data from this quadrant, if the area that we need falls completely within the valid area of the node, then we do not need to load any additional data. If the area that we need is not fully enclosed by the valid area, then we load the missing part and increase the valid area of the node accordingly.

A typical dataset would contain several tables representing different layers of the map. While each layer is stored in a separate PMR quadtree, there is only a single binary heap data structure for all the layers combined. This way, when a user stops working with one of the layers, its data will be automatically and gradually removed from the cache and will be replaced with the data needed currently.

As the user explores the content of the database using a graphical viewer, s/he is basically retrieving all the objects stored in the database that overlap the current viewing window. When the content of the spatial structure overlapping a certain query object needs to be drawn, a tree traversal is performed to find all the objects in the PMR quadtree that overlap the query object. At times, we find that the internal nodes are "invalid" which means that either they were not loaded yet or they were previously removed by the memory management process when they were not used for some time. In such a case, the data needs to be reloaded.

The algorithm contains two steps. In the first step, the system finds out what areas need to be loaded from the server and builds a collection of rectangles that represent this area.

In the second step, the algorithm takes the list of rectangles returned by the first step and loads all the data from the server that lie within the area defined by this collection of rectangles. Next, for each rectangle loaded, it adjusts the corresponding PMR node status.

Now, when we need to display all data that overlaps a given window w, we can look at not just the valid/invalid identifier of each PMR node that overlaps w, but instead we can also check the *validSubarea* field of the invalid nodes. If the intersection of the window w with the PMR block is fully contained in the node's *validSubarea*, then we know that all the necessary data for this window query is already in the database, even if the PMR node is not loaded fully. When the drawing function is called, it already knows that all the data is already loaded in the memory and it simply steps through the overlapping PMR nodes and displays their contents.

An obvious limitation of the memory-only approach is the maximum amount of space that can be utilized for local data storage. This approach is the only one available when the client runs on a platform that has no secondary memory (e.g., disks) available. Such an environment is usually present on smaller handheld devices or on Java applet-based viewers. Various SQL-based DBMSs exist for many platforms that can be used to facilitate local caching using available disk space.

4 Utilizing Auxiliary Servers

Development of Internet technologies has introduced various methods for utilization of additional servers to improve performance for end users who connect to external servers. One of the first and most popular methods is caching. Caching can be implemented directly within the end user's browser, or it can also be implemented within the user's network, on the gateway (proxy) between the network and the outside Internet. In the latter case, the same cache can be shared among several users.

Obviously, the rationale for introducing these proxy servers between the client and the Internet is that the responsiveness of the proxy server with respect to the end user's browser is much higher than if the data was requested directly from the original host. This is due to the higher network speed between the client and the proxy server compared to the network speed between the client and the original host. Another factor can possibly be the lower load and higher responsiveness of the proxy server since it only handles traffic for a few users and therefore can process requests more efficiently.

An example deployment of a proxy server is an emergency situation illustrated in Figure 1.2. There, multiple first responders equipped with handheld devices link with a mobile communication van or similar vehicle. This vehicle is equipped with a wireless router as well as with satellite or similar communication technology and facilitates connectivity with the central computing facilities.

4.1 Static Proxy

In some cases, the main spatial server provider and the individual users of this database are from within the same organization or these organizations collaborate closely. If this is the case and the spatial data is rather static (i.e., updates in the database are not performed frequently), it may be feasible to execute a one-time step of copying all the spatial data stored in the main spatial database onto the auxiliary database running on the proxy server. In such a scenario, the auxiliary database needs to be preloaded with the spatial data from the central SAND server when the system is being installed as well as possibly periodically after that[1]. The frequency would depend on how often the data on the central server changes. This approach is especially effective if updates on the central spatial server are performed in regular intervals rather than dynamically. For instance, a new data set may be released once a month or once a year instead of applying partial updates continuously.

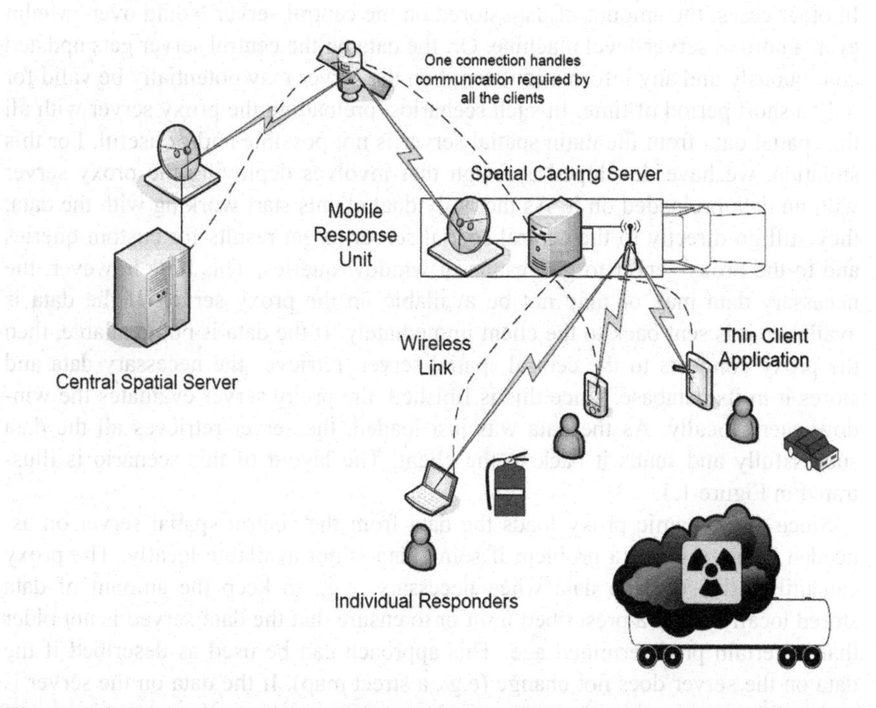

Figure 1. 2. Emergency response service deployed a mobile unit (e.g., a mobile van) in support of the operations. This unit can cover the area with a fast wireless network access and provide a proxy service for spatial operations. Individual responders can utilize the applications on their mobile devices more efficiently.

[1] This arrangement is similar to setting up a mirror server. The difference is that a mirror server is typically a copy of the primary server and can provide any functionality that the primary server does. In this case, the proxy only stores spatial data of the background map and facilitates window queries. The central server is still used to evaluate complex custom queries as initiated by the user.

Since the complete valid map resides on the proxy server, there is no need for the client to ever connect to the central spatial server for window queries. There is also no need for the proxy server to talk to the spatial server, to receive updates or for any other reason. Therefore, the only traffic generated by this scheme involves the SAND Internet Browser clients communicating with both the central spatial server (e.g., SAND server) and the auxiliary proxy server.

4.2 Dynamic Proxy

In other cases, the amount of data stored on the central server would over- whelm even a normal server-level machine. Or, the data on the central server gets updated continuously and any information stored on the server may potentially be valid for only a short period of time. In such scenarios, preloading the proxy server with all the spatial data from the main spatial server is not possible and/or useful. For this situation, we have developed a design that involves deploying the proxy server with no data preloaded on it. As the individual clients start working with the data, they still go directly to the central spatial server to get results for custom queries and to the proxy server to get results of window queries. This time however, the necessary data may or may not be available on the proxy server. If the data is available, it is sent back to the client immediately. If the data is not available, then the proxy connects to the central spatial server, retrieves the necessary data and stores it in its database. Once this is finished, the proxy server evaluates the window query locally. As the data was just loaded, the server retrieves all the data successfully and sends it back to the client. The layout of this scenario is illustrated in Figure 1.3.

Since the dynamic proxy loads the data from the central spatial server on as-needed basis, it is not a problem if some data is not available locally. The proxy can utilize this to drop data when necessary, e.g., to keep the amount of data stored locally under a prescribed limit or to ensure that the data served is not older than a certain predetermined age. This approach can be used as described if the data on the server does not change (e.g., a street map). If the data on the server is updated frequently, then the server needs to notify its clients that a certain part of the database was updated. In response, the clients drop the corresponding data from their cache and will reload it the next time a user requests it.

4.3 Implementation Details

The SAND Internet Browser running on clients is implemented in Java and its connection with the external servers is facilitated via Java Database Connectors (JDBC) modules provided by the respective database vendors.

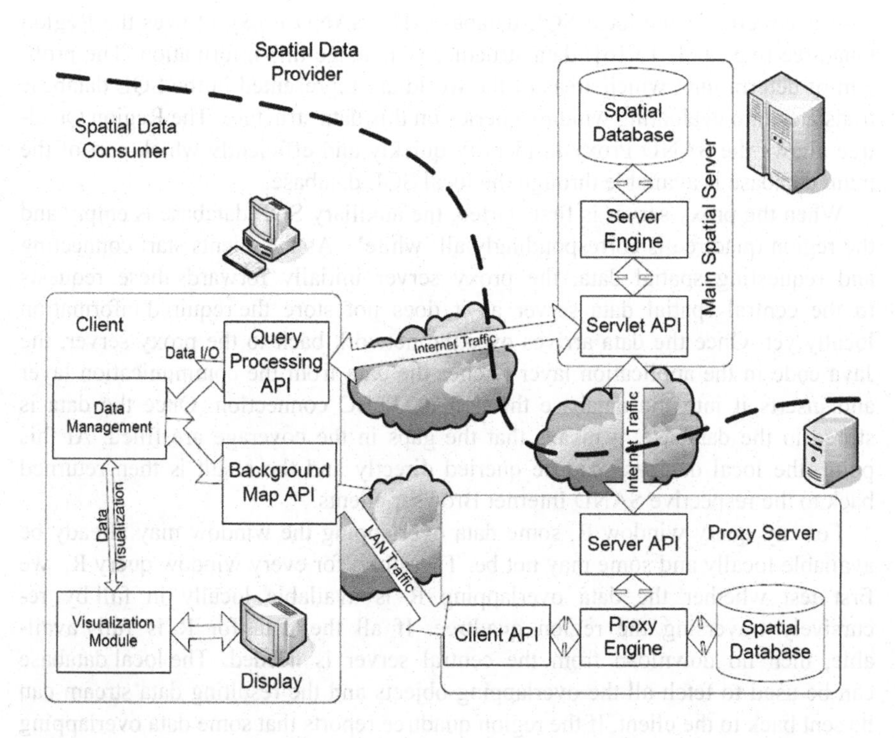

Figure 1. 3. Dynamic Proxy — The proxy server is installed with no data on it initially. It connects to the central spatial server and if a request comes from a client for data not available locally, the proxy retrieves the data from the central server, caches it locally, and sends it back to the client.

The SAND Proxy, the implementation of the proxy server outlined in general above, is a combination of two modules. The first one is an off-the-shelf SQL database[2] responsible for storage of spatial data storage used in handling of window queries. Note however, that the SQL database does not have any information regarding what data it contains compared to the content of the main spatial server. This is the responsibility of the second module, it maintains information about which parts of the SQL database are currently valid (i.e., which parts fully mirror the content of the central SAND database). Additionally, it facilitates communication the clients and, in case of the dynamic proxy, with the SAND server that performs the role of the central spatial database.

The second module in essence implements a second database which maintains the information about which area of the "world" that is stored in the central data-

[2] The SAND Internet Browser system has been used with MySQL [4] and PostgreSQL [9] but other SQL databases could also be plugged in.

base is covered in the local SQL database. The SAND Proxy utilizes the Region Quadtree (e.g., [14, 15,16]) data structure to manage this information. The problem of determining which areas of the world are represented in the SQL database translates into evaluating window queries on this data structure. The Region Quadtree allows the SAND Proxy to identify quickly and efficiently which part of the main database is available through the local SQL database.

When the proxy server is first started, the auxiliary SQL database is empty and the region quadtree is correspondingly all 'white'. As the clients start connecting and requesting spatial data, the proxy server initially forwards these requests to the central spatial data server as it does not store the required information locally yet. Once the data arrives over the network back to the proxy server, the Java code in the application layer fetches the data from the communication layer and inserts it into the database through its JDBC connection. Once the data is stored in the database, it means that the gaps in the coverage are filled. At this point, the local database can be queried directly and the result is then returned back to the respective SAND Internet Browser clients.

For any query window R, some data overlapping the window may already be available locally and some may not be. Therefore, for every window query R, we first test whether the data overlapping R is available locally in full by recursively traversing the region quadtree. If all the data for R is fully available, then no download from the central server is needed. The local database can be used to fetch all the overlapping objects and the resulting data stream can be sent back to the client. If the region quadtree reports that some data overlapping R is missing in the local database, then a download of all the data overlapping R in its entirety is requested from the server. While it will re-load some data that are already present locally, the benefit is that the overhead is much smaller, as only a single window query is submitted to the central server. Any would-be duplicates are ignored by the SQL databases as the data table structure is set up to enforce uniqueness of individual objects stored. This ensures that we do not store duplicate entries in the cache. The decision to aggregate multiple smaller queries into a single larger one is one of the aspects of our design.

After the data overlapping R is loaded from the server, the region quadtree is updated to mark R as fully loaded. This is done through top-to-bottom insertion into the region quadtree — that is, by recursively visiting all overlapping nodes, marking them as covered if they are fully overlapped. Or, in case of a partial overlap and unless the maximum depth was reached, the node is subdivided into four children and the same operation is performed recursively. If there is still just partial overlap of R and leaf node N when the algorithm reaches the maximum allowed decomposition level, then we mark the node as covered. This ensures that any subsequent window query that is simply a result of a lateral movement (i.e., a scroll operation) along the same axis as the window edge that intersects N won't report missing data due to the same N and cause another download request to the central server. Of course, the drawback is that the region tree reports N as available in the SQL database while part of the data overlapping N is in fact missing. In reality, this area is very small (a fraction of the node on the region quadtree maximum depth level) and will typically be loaded before the data is needed — once the window R moves such that it overlaps N in full. This is because N's empty neighbors will trigger download of data overlapping R thus filling the gap in N's coverage as well.

This approach guarantees that the proxy is always able to provide the data requested by the client, while efficiently caching the data for future use. While this approach as described, assumes the auxiliary database has enough resources to store all data that flows through the proxy, it is not a requirement. If the availability of sufficient resources cannot be guaranteed, the same method used in Section 3.2 that allows for a limited amount of storage space can be applied here as well.

5 Building Combined Solutions

This section describes how the individual building blocks presented previously can be combined together to build a complete spatial database visualization solution. Results of experiments are given that provide guidelines for selection of appropriate designs given specific deployment scenarios.

5.1 Modular Design and Chaining

While different host types may be used to cache spatial data, their functionality is similar. Their goal is to store the data that have passed through up to their efficient capacity. The individual proxy modules can be stacked on top of each other, where the node closest to the actual displaying client has the smallest capacity and usually stores a subset of data of its successor in the chain. The farther up in the chain that we go from the client, the more data and processing power the node within the chain has.

This is because when a client requires a certain data range and cannot find this information locally, it sends the request to the next cache/proxy node.If the data is available there, then it is served. If it's not available there, then the cache/proxy requests the same data farther up the chain. This process repeats until the data is reached, in the worst case in the main spatial data server. Once the data is reached, it is sent back the same way the requests came, i.e., all caches/proxies on the way between the client and the successful data repository will get the chance to store the data as well. Since the layers closer to the client have typically smaller capacity, they would usually have to drop some of the data first and thus end up storing subsets of data available on the proxy. This proxy hierarchy is outlined in Figure 1.4. Of course, what data can be expected to be stored on the proxy becomes less clear once the proxy serves multiple clients. In such a case, the proxy may get overwhelmed by requests from another client in such a way that it is forced to drop all data loaded for our client. In this case, our client may still hold some data while the proxy no longer does.

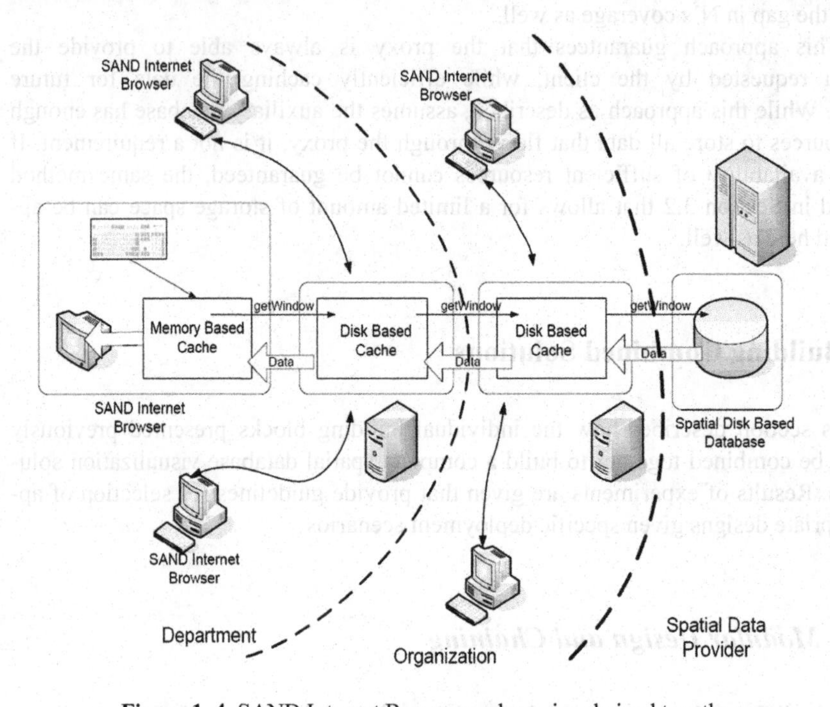

Figure 1. 4. SAND Internet Browser and proxies chained together

Regardless of the type of platform managing the data, the data is always stored in a spatial data structure (e.g., some variant of a quadtree). The main data server runs a full-blown spatially-enabled DBMS. The proxies and clients however only perform a subset of operations of a normal DBMS in order to support the limited functionality required by this layered system of caches/proxies.

This layered system is only used for base map visualization. It is not used to evaluate queries. The common interface for nodes participating in the stacked caching system turns out to be very simple:

implements: getArea(Rectangle area)
requires: getArea(Rectangle area)

This means that each participant in the infrastructure must be able to perform a remote procedure call (RPC) representing a window query on its parent within the hierarchy (where the parent means the node closer to the main server).

It also needs to provide a window query interface, i.e., allow nodes closer to the client to submit window queries (RPCs) to it.

Above, we have shown that individual computing platforms can be linked together to create a chain of caching proxies that link the client's visualization module with the central spatial database. Not all computers within this chain need to employ the same caching method. They only need to implement the above interface. The actual implementation can vary depending on the hardware parameters of that platform as well as other factors. However, even within each computer, the individual caching methods do not need to be used in an isolated fashion. The caching concept can be generalized to involve an arbitrary number of caching layers that can be stacked on top of each other in the order of the speed with which they are able to serve the content. Many times, the speed of delivery is inversely proportional to the volume of data any given layer can store efficiently or at all. For instance, access to data stored in primary memory is fast but the storage capacity is limited. On the other hand, a disk-based memory has substantially larger capacity but access to the data is slower.

Note that accessing the central data server can be considered to be within the framework of such a layer as well, and it would be the last and slowest layer; however, it always succeeds (never generates a page fault). So we see that it does not matter whether the data served by any given layer is stored locally or in a remote location. Thus, this concept allows us to generalize the caching into multi-server setups, or even to a peer-to-peer environment. All the client needs to know is in which order it should turn to individual data providing layers. Note that the border between data cache and data server is fuzzy as individual clients can share caches on servers closer to them than the original server, in which case such caches would actually serve as sort of proxies in such environment.

6 Evaluation

Our research explores the impact of various types of techniques for chaining different caching layers together on the performance of the solution. We investigated different scenarios and suggest ideal combinations of caching based on the types of devices used, usage model (e.g., number of users looking at the same data), network speed, and other factors.

Specifically, we have designed and implemented the following caching methods and investigated properties of the SAND system created by chaining them in various combinations:

- Client

  1. direct access — client communicates directly with main spatial server with no local caching

2. local caching — client caches data in its memory

- Proxy
  1. pre-loaded data — the local SQL database is pre-loaded with all spatial data from the server
  2. dynamically-loaded data — the local SQL database is loaded dynamically based on the requests coming from the clients

The behavior of the whole system depends on a number of factors, many outside our reach (e.g., the network latency, number of concurrent users, or even the exact implementation of the garbage-collection algorithm in the underlying operating system or virtual machine, etc.). This also makes a rigorous comparison with other existing systems that aim to serve the same goal (e.g., MapQuest) difficult as we are not able to run performance tests of both systems in identical environments. Therefore, the nature of the SAND system and a MapQuest-type system makes their comparison difficult. Of course, we have tried to minimize the impact of external factors. This is achieved by utilizing the same hardware and software platforms for both systems, the same networking environment as well as identical data sets, queries or sequences of queries. In addition, the parameters of the server platform, the networking environment, and the type of datasets and queries that were run on them were chosen to be typical for the types of deployments that we suggest would benefit from this system.

The goal of this evaluation is not necessarily to determine that one approach is better in every scenario. Instead, we aim to identify what approach is the best one for different methods of deployments and provide the system administrator and user with guidelines for selecting a solution best suitable for their specific needs. Besides comparing vector-based SAND Internet Browser against a bitmap solution, we also deployed SAND in several different ways utilizing its modularity as described in Section 5.1.

6.1 Comparison with Raster-Based Visualization

For our performance evaluation, we used TIGER datasets from the U.S. Census, specifically the street maps for states in the Mid-Atlantic region. This includes all the roads and streets in Virginia, Maryland, District of Columbia, New Jersey, and Pennsylvania. There are over 7,500,000 entries in this combined dataset. Each entry corresponds to a single line segment, where each actual street may be represented by one or more line segments in the map. The total size of the data stored in the format distributed by U.S. Census is over 700MB.

Our performance testing aims to compare different methods of deploying SAND's vector-based approach to remote mapping with the bitmap based approach employed by such popular systems such as MapQuest. In order to run both systems in the same environment, we chose MapServer [2] to represent the bitmap approach. This allows us to deploy both systems on the same hardware, using the same operating system and within the same networking environment. This also enabled us to minimize performance differences caused by factors that are not directly related to the design of spatial data management.

6.2 Typical Usage Scenarios

A user of a mapping or GIS system frequently performs the following operations while navigating the map:

- Zoom in — view an area of interest in more detail.
- Fast Scroll — move the viewable area to the left and to the right, or up and down by large increments. In our scenario, the map moves by one half of the window size, i.e., there is 50% overlap between the old and new views.
- Fine Scroll — move the viewable area to the left and to the right or up and down by small increments, perhaps only by a fraction of the window width or height. In our scenario, the map moves by 10% of the window size, i.e., there is 90% overlap between the old and new views.
- Zoom out — view a larger area of the map within the viewable window.

We expect (and confirm our expectations by running experiments) that the cost of each visualization operation (zoom, pan) for the MapServer approach will be approximately constant given a constant data density (i.e., the number of objects to be visualized for a fixed view area size) and viewable area size. If the number of elements within the viewable area remains the same, then the cost of the spatial query and the cost of subsequent rendering and bitmap transfer remains the same as well. Thus we see that when the number of objects visible as a result of a visualization operation remains the same, the cost of updating remains constant as well. Given a server platform, the MapServer system responsiveness will depend on the network speed and latency. The situation for the SAND Internet Browser is different. There, the system takes a more complex approach when processing visualization requests and the response time depends on the nature of the request as well as on the history of similar requests preceding this one.

As mentioned above, we have selected several typical operations that users of a mapping system or GIS would perform most often while navigating around the map. These operations include zooming in and out and panning/scrolling. First, we compare MapServer with the standard SAND Internet Browser setup that only involves the central data server and the SAND Internet

Browser client. Later we also compare MapServer with a deployment of the SAND Internet Browser in an environment where data cannot be stored locally. We conclude with a comparison of the estimated performance of the tile method as typified by Google Maps in the same environment in which both MapServer and the SAND Internet Browser were deployed.

For the SAND Internet Browser, we measure the execution time in two scenarios:

- The data to be visualized as a result of the user's operation is already cached on the system.
- The data to be visualized as a result of the user's operation is not yet cached on the system and has to be loaded dynamically from the server.

For MapServer, the bitmap is always downloaded from the server for each new operation.

To measure performance across various deployment scenarios (here represented by different properties of the network connection), we emulate networking environments that correspond to several typical methods of achieving connectivity (i.e., "hookup") on mobile devices as well as fixed workstations. Table 1.1 describes these connections.

Table 1. 1. Properties of various network connection (i.e., "hookup") methods

Connectivity (i.e., "hookup") methods		
Hookup	Bandwidth (kB/sec)	Delay (sec)
Modem	7	0.3
Broadband	183	0.2
Satellite	62	1
LAN	1,250	0.002

To emulate standard usage scenarios, all TCP/IP parameters of the networking layer were left at their default values even though for some types of connectivity adjustments of these parameters may improve the overall throughput.

To emulate different networking properties in our test environment, we have utilized NIST Net [6], a general-purpose tool for emulating performance characteristics in IP networks. We have configured NIST Net using networking parameters typical for individual connectivity methods (Table 1.1) to measure the performance of the SAND system in different deployment scenarios.

For the pure client-server environment (i.e., no auxiliary servers), the performance was tested for the following three basic client-server architecture states. First, the *cached* SAND Internet Browser state refers to a scenario where the SAND Internet Browser provides local caching and the data to be displayed as a response to the sequence of scroll operations is already available in the client's memory. Second, the *direct* SAND Internet Browser state refers to a scenario where the client does not cache data locally and downloads all the data from its server. This represents the pure client-server setup where the client communicates directly with the central server. Finally, the *dynamic* SAND Internet Browser state refers to a scenario where the client provides local caching but the necessary data is not available in the local cache yet.

Results of a performance comparison of MapServer with the SAND Internet Browser for fine scrolling can be seen in Table 1.2. During a sequence of fine-scroll operations, the previous window overlaps of the next window 90% of the window area. This means that the SAND Internet Browser can use a fast bitmap copy operation to transfer the part that can be reused to an other location of the screen and it needs to rasterize only 10% of the window using vector data stored either locally or downloaded from the server. We see that the performance in case of cached data is essentially the same across all hookup methods. This is because all data is cached and no data needs to be transferred across the network. Slight differences are due to operations performed by unrelated background processes (e.g., Java VM garbage collection). Also note that while the direct approach does not use any caching and loads all data from its upstream provider all the time, its data management overhead is lower. Hence, for faster types of network connections the direct method tends to perform better, while the cached methods is typically better for slower connection methods.

Table 1. 2. Performance comparison of MapServer, the SAND Internet Browser, and the Tile Method (e.g., Google Maps) for fine scroll. The table indicates the time in seconds to perform 20 subsequent fine-scroll operations.

Fine scrolling/Local Panning					
	Cashed	Noncached			
Hookup	SAND Internet Browser			Map Server	Est. Tile Method
	Cashed	Direct	Dynamic		
Modem	6.6	124	80	179	18
Broadband	6	20	38	52	5
Satellite	5	81	85	181	18
LAN	5	10	33	18	2

Results of a performance comparison of MapServer with the SAND Internet Browser for zooming in can be seen in Table 1.3. The starting viewable window showed 25,000 line segments and each zoom-in operation doubled the map scale, i.e., both the x and y coordinate ranges were halved. Thus, the area before the zoom-in operation is four times as large as the area displayed after the zoom-in operation. We measured the time it took to execute five subsequent consecutive zoom-in operations with the last view showing only dozens of line segments.

Note that the viewable area resulting from the zoom-in operation is always a subset of the viewable area that existed prior to the zoom-in operation. Thus, for the caching SAND Internet Browser, the data to be displayed after any zoom-in operation will always be available in the cache. Here we assume that the client uses the same data set on all the zoom levels involved. In practice, zooming in may require the client to display data from additional data layers which may not be available in the cache yet.

Table 1. 3. Performance comparison of MapServer, the SAND Internet Browser, and the tile method (e.g., Google Maps) for zoom-in. The table indicates the time in seconds to perform five subsequent zoom-in operations. The results for the dynamic SAND Internet Browser method are not applicable (N/A) since the data will always be cached from the previous operation (assuming all zoom levels retrieve data from the same dataset).

Zoom In					
Hookup	Cashed	Noncached			
	SAND Internet Browser			Map Server	Est. Tile Method
	Cashed	Direct	Dynamic		
Modem	0.5	10	N/A	44	44
Broadband	0.8	3	N/A	12	12
Satellite	0.5	10	N/A	44	44
LAN	0.8	1	N/A	5	5

Results of a performance comparison of MapServer and the SAND Internet Browser for the zoom-out operation can be seen in Table 1.4. This query test is essentially a reverse of the zoom-in operation with a single important distinction in the caching SAND Internet Browser. In particular, while each zoom-in operation can expect to have all the necessary data cached from the previous step, in the zoom-out operation this is not necessarily the case. Consider a scenario when the user moves around in a zoomed-in (e.g., street) level and then tries to zoom out (e.g., to city level). As the viewable area grows, not all the data objects that overlap this area are necessarily cached.

For the zoom-out operation, the starting viewable window showed a large detail containing only a few dozens of line segments. Each zoom-out operation expands both the x and y coordinate ranges twice. Thus, the displayed area before the zoom-out operation is four times smaller than the displayed area showing after the zoom-out operation. We measured the time it took to execute five subsequent zoom-out operations, the last view was showing about 25,000 line segments. We considered both scenarios outlined above for the SAND Internet Browser. One scenario captures the situation where the data to be shown after the zoom-out operation is already in the cache (i.e., the zoom-out operation was preceded by a zoom-in operation without any panning operations in between). The other scenario explores a situation when the data to be shown after the zoom-out operation is not in the cache and has to be fetched from the spatial server.

Table 1. 4. Performance comparison of MapServer, the SAND Internet Browser and the tile method (e.g., Google Maps) for zoom out. The table indicates the time in seconds it took to perform five subsequent zoom-out operations.

Zoom Out					
Hookup	Cashed	Noncached			
	SAND Internet Browser			Map Server	Est. Tile Method
	Cashed	Direct	Dynamic		
Modem	1.8	26	48	45	45
Broadband	1.6	5	22	12	12
Satellite	3.2	17	36	45	45
LAN	2.3	2	20	5	5

Table 1. 5. Performance comparison of MapServer, the SAND Internet Browser, and the tile method (e.g., Google Maps) for fast scroll (global panning). The table indicates the time in seconds to perform 20 subsequent fast-scroll operations.

Fast Scrolling/Global Panning					
Hookup	Cashed	Noncached			
	SAND Internet Browser			Map Server	Est. Tile Method
	Cashed	Direct	Dynamic		
Modem	3.9	108	109	161	80
Broadband	3.9	19	54	44	22
Satellite	3.9	80	104	165	82
LAN	3.8	9	48	14	7

Table 1.5 shows the results of a performance comparison between MapServer and the SAND Internet Browser for global panning. Unlike in the Local Panning/Fine Scrolling scenario evaluated above, in the global panning operation, a

large portion of the post-panning viewable area does not overlap the prepanning viewable area. This means that the SAND Internet Browser must load a large portion of the new viewable area from the locally cached data or from the central spatial server. Given this realization, we again measure the performance of the SAND Internet Browser for two distinct scenarios:

- The data to be visualized as a result of the user's operation is already cached on the system.
- The data to be visualized as a result of the user's operation is not yet cached on the system and has to be loaded dynamically from the server.

MapServer, as always, generates a new bitmap on the server and pushes it onto the client. Each view was showing about 25,000 line segments during this panning operation. As we can see, each of the tests performed above repeats the same operation under the same conditions. This provides us with a comparison of each possible operation under given conditions (in terms of network parameters) separately. While in a real life deployment the network parameters will likely remain fixed during each session, the sequence of operations will probably be a combination of the available operations. In other words, the user will probably not use solely the fine-scroll or the zoom operations, instead they would typically do some scrolling, then zoom in, scroll some more, zoom out, etc. The typical sequence structure and duration of such a session depends on the nature of the scenario. Reviewing a larger area for certain properties may involve much scrolling and a minimum of zooming. Investigation of multiple separate locations may involve more zooming in and out with a minimum amount of panning.

The user will rarely work under conditions when the spatial data is either fully cached all the time or not cached at all in any step. Depending on the exact usage patterns, the user can expect to benefit from the caching for some portion of his or her operations. The success rate of the caching mechanism will depend on numerous factors. The first is the time at which the operation is executed. The cache will be empty right after the start-up of the client application. So the user can expect to be fetching data from the server for most such operations initially. Thus, the initial performance of the caching SAND Internet Browser will appear close to what we have shown above under the non-cached data columns (i.e., direct or dynamic). Once the cache is filled with data, the success rate will depend on the extent to which the user's spatial operations are localized. If the user visualizes information directly within the same limited area (e.g., fine scroll or zoom in), then most of the operations will use the cached data. In such a scenario, the performance will be close to what we have shown above in the cached data column. Most of the time the sequence of operations generated by the user will trigger a mixture of cached and non-cached data retrievals. Thus, we can consider our cached and non-cached results as the extreme cases of what a user may expect and a typical experience lies somewhere in between.

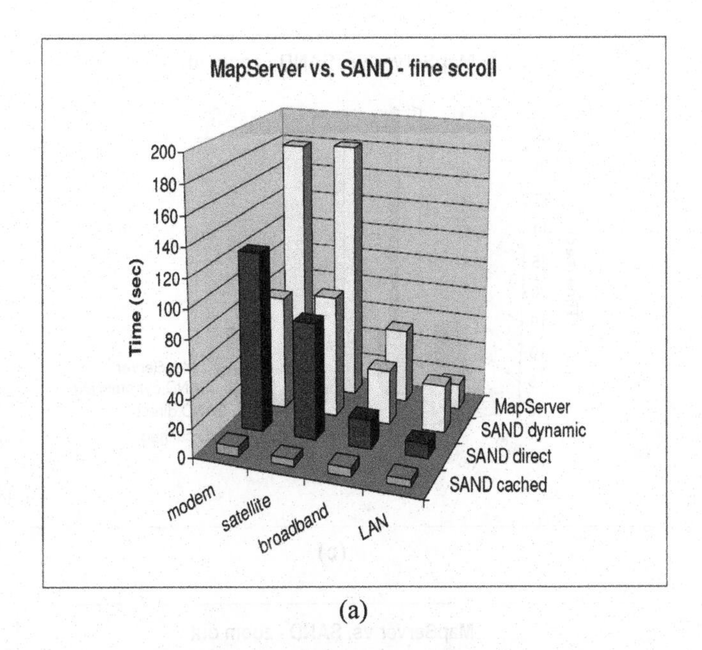

(a)

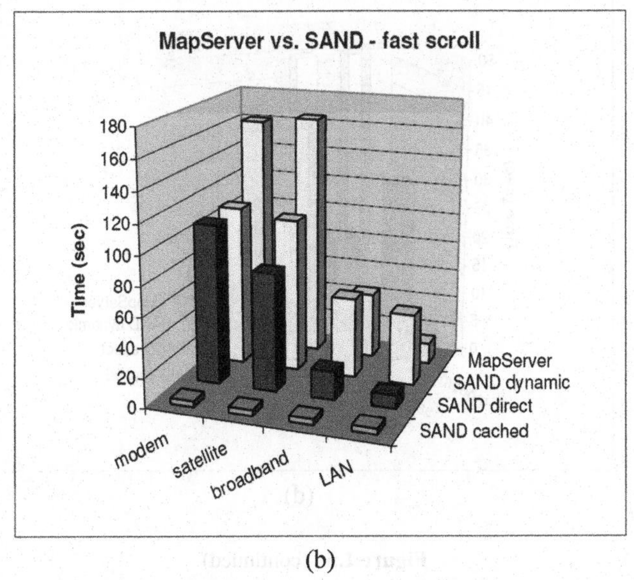

(b)

Figure 1. 5. Comparison of a bitmap (MapServer) approach with the vector-based SAND approach for remote spatial data visualization. Assuming all zoom levels retrieve data from the same dataset, note that the SAND dynamic scenario for the zoom-in operation is not applicable as all data is already cached from the previous view (denoted by 'X').

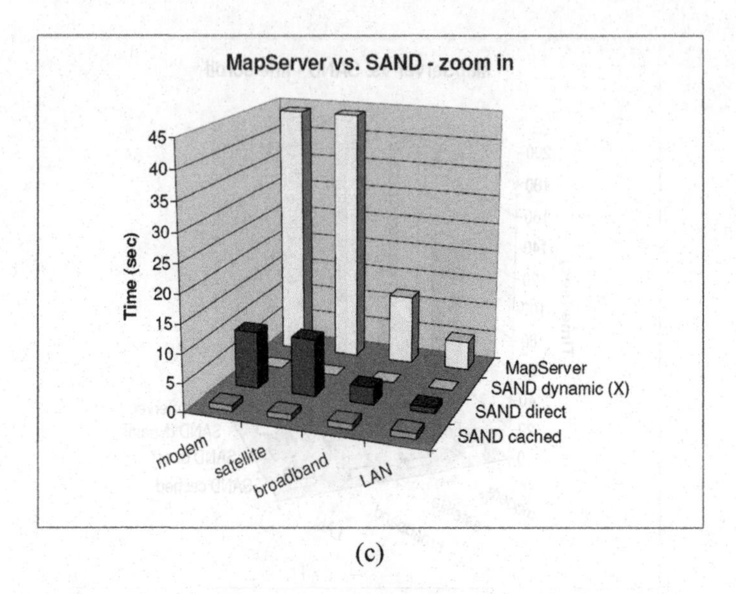

(c)

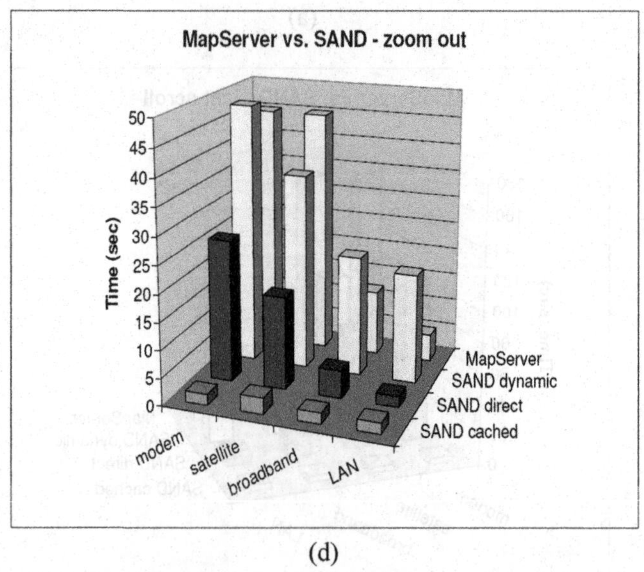

(d)

Figure 1.5. (continued)

Figures 1.5a–1.5d display Tables 1.2–1.5 graphically. The figures show that in most deployment scenarios, network environments, and usage patterns, the user can expect to have a substantially better experience when using the SAND Internet Browser than when using a pure bitmap system.

6.3 Performance Comparisons for Deployments Utilizing Auxiliary Servers

In the previous section we compared the SAND Internet Browser-based system that involved a caching and non-caching client and a central spatial server with a bitmap based system represented by MapServer. Here, we evaluate a scenario outlined in Section 4 where a small footprint wireless-capable handheld devices (e.g., smart/cell phones, PDAs and other similar devices) not capable of storing data locally can be used within the SAND Internet Browser-based architecture. Note that the bitmap approach is still valid as the client does not store any data locally and thus this method is still applicable even on these mobile devices.

We examine three different deployment scenarios. The first scenario involves the static proxy (section 4.1) and the state is termed *preloaded.* The remaining two scenarios involve the dynamic proxy (section 4.2). The first of the two assumes that the user just started the application so that no cached data is available yet. We call this state *clean.* Since the proxy server can provide its services to multiple users, we also assume that this user is the first one to request this particular data. The second dynamic proxy scenario assumes that the same data was already accessed before (by this or another user), and thus it is already available on the proxy server. This state is termed *cached.*

Using the auxiliary server deployment example from section 4 where first responders use handheld devices to communicate with the central facilities via a mobile van, we see that the communication link consists of two parts.The first link connects the devices and the mobile van, while the second link connects the van with the central facilities. We presume that in emergency scenarios such as these, the connectivity between the handheld devices and the van is faster than the connectivity between the van and the central facility. Based on the assumptions for such an emergency response deployment, we assume that the mobile teams will be able to connect to the central facilities over a satellite link. Locally, the connection between the individual response team members will be wireless (e.g., WLAN 802.11b/g). This emulation is again facilitated by using the NIST Net tool.

Table 1.6 shows the results for different usage scenarios that involve auxiliary servers. The client to auxiliary server link is of a wireless LAN type. The link between the auxiliary server and the central spatial server is a satellite connection. Figure 1.6 shows the performance of a system that utilizes auxiliary servers. As we see, MapServer performs better when compared to the SAND Internet Browser on a freshly installed system where no data has been pushed through the infrastructure yet (labeled “clean” in the figure) and the proxy server cache is still empty. This is because of the additional overhead of copying the necessary data from the central server, a step that MapServer completely bypasses. Once the cache is loaded with data, we see that the SAND Internet Browser performs at least as well as, and, most of the time, significantly better than MapServer. If the auxiliary server is preloaded with the data, then the improvement in the performance of the SAND Internet Browser over MapServer is even more pronounced. Note that for the zoom-out operation, the preloaded approach is substantially faster than the cached approach. While there is no network traffic in either case, the cached method has extra overhead (e.g., before sending the data to the client it first needs to verify if any additional data needs to be downloaded from the central server).

Table 1. 6 Performance comparison of various operations for MapServer and the SAND Internet Browser using the auxiliary server deployment method. The scroll operation values represent the time (in seconds) it took the system to process 20 subsequent scroll operations. The values associated with the zoom operations indicate the number of seconds it took the system to process five consecutive zoom operations. The result for the zoom-in operation in the dynamic SAND Internet Browser method is not applicable (N/A) since the data will always be cached from the previous operation (assuming all zoom levels retrieve data from the same dataset).

Auxiliary Server-based Deployment				
Operation	MapServer	Proxy SAND Internet Browser		
		Preloaded	Dynamic	
			clean	cached
Fast Scroll	165	19	568	52
Fine Scroll	181	29	520	75
Zoom In	44	2	N/A	12
Zoom Out	45	6	58	48

6.4 Comparison with the Tile Method

The tile method, another bitmap-based method providing an alternative to the MapQuest-type bitmap approach, was described in Section 2. While we cannot run formal experiments comparing Google Maps or MS Virtual Earth with the SAND Internet Browser directly, we can estimate what their performance would be within the same environment in which MapServer and the SAND Internet Browser were deployed. Assuming that the cost of generating the tiles on the server is negligible, the determinative factor for the cost is the amount of data sent from the server to the client. Since the tile method allows for tile reuse, only the newly visible areas will trigger further download.

When all the necessary data is fully cached from the previous steps, the response times for the SAND Internet Browser and the tile method are essentially instantaneous. For the tile method, the browser simply needs to redisplay the cached tile images which takes no time. Similarly, the SAND Internet Browser also just needs to render and display the cached vector data, done by retrieval and rasterization, which also only takes a fraction of a second (e.g., Table 1.2 once the execution time is divided by 20, the number of scroll operations, to yield the time per fine-scroll operation). Hence, for scenarios where data is fully locally cached, we consider the performance of the tile method and the SAND Internet Browser to be comparable.

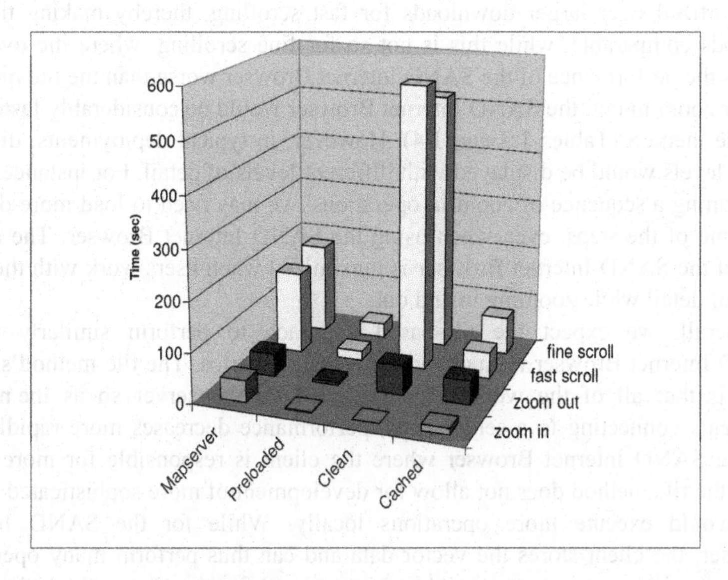

Figure 1. 6. Performance comparison of various operations for MapServer and the SAND Internet Browser using the auxiliary server deployment method. Note that the non-cached (clean) scenario for the zoom-in operation is inapplicable.

Comparing the tile method with MapServer, we find that as the fast-scroll operation reuses 50% of the visible area, the tile method would be twice as fast as MapServer (see Table 1.5. The fine-scroll operation reuses 90% of the area, so the tile applications would be about ten times faster than MapServer (see Table 1.2). Both tables indicate the time in seconds it took to perform 20 subsequent scroll operations. Using the tile method, zoom in/out (if offering the view for the first time and thus not using cached data) would take as long as MapServer as the specific bitmaps are not yet available on the client and thus they have to be loaded from the server in full—that is, they must be reused. Therefore, MapServer results for zoom in/out shown in Tables 1.3 and 1.4 also apply as estimated results of the tile method.

Comparing the tile method with the SAND Internet Browser we find that for fine scrolling (see Table 1.2) the tile method would be faster than the SAND Internet Browser when the data is not already cached (i.e., direct and dynamic). For fast scrolling (see Table 1.5), the tile method would still be slightly faster or perform comparably when the data is not already cached (i.e., direct). The dynamic case is slower in the SAND Internet Browser due to overhead in setting up the caching such as traversing the PMR quadtree, etc. The rationale for the comparable behavior for fast scrolling is that, overall, there is a fixed per-update overhead involved in requesting, handling and storing the data which is higher for the SAND Internet Browser due to the vector format of the data. This overhead is amortized over larger downloads for fast scrolling, thereby making the two methods comparable, while this is not so for fine scrolling, where the overhead makes the performance of the SAND Internet Browser worse than the tile method.

For zoom in/out, the SAND Internet Browser would be considerably faster than the tile method (Tables 1.3 and 1.4). However, in typical deployments, different zoom levels would be displayed with different levels of detail. For instance, when performing a sequence of zoom-in operations, we may need to load more data after some of the steps, even when using the SAND Internet Browser. The advantage of the SAND Internet Browser is maximized when users work with the same level of detail while zooming in and out.

Overall, we expect the tile-based approach to perform similarly to the SAND Internet Browser in many actual usage scenarios. The tile method's drawback is that all of the work is concentrated on the server so as the number of clients connecting to a server rises, performance decreases more rapidly than for the SAND Internet Browser where the client is responsible for more work. Also, the tile method does not allow for development of more sophisticated clients that would execute more operations locally. While for the SAND Internet Browser, the client stores the vector data and can thus perform many operations (such as window or nearest neighbor operations);for the tile method, the client only has access to the bitmap tiles, which do not provide data for such localized operations. So, we see that the SAND Internet Browser is a better platform for developing smarter, more independent client applications.

7 Conclusions and Future Research

We presented a new vector-based system for remote access to spatial databases that could be used where the traditional raster-based approaches do not work too well. We compared the performance of a bitmap raster-based system with the vector-based SAND Internet Browser system. Our experiments allow us to suggest the best type of a remote spatial data visualization tool for a given deployment scenario.

We have developed a modular design for the infrastructure that facilitates remote spatial data access. We applied it to realize several specific types of the SAND Internet Browser system deployment. The best-performing deployment depends on the environment in which the system is to be used. Generally, the system can either be deployed so that the clients communicate directly with the central spatial server. Alternatively, in situations where the client runs on a thin platform or where the service is shared among several co-located clients, an auxiliary server could be used to improve the overall solution's performance.

Future research directions include investigating methods for caching frequently used data in the form of bitmap tiles instead of vectors. While these tiles would only be usable in given views (in terms of zoom factor and layers displayed), they would also allow skipping of repeated rasterization steps. The result would be a hybrid between the SAND Internet Browser and the tile method used by Google Maps and Microsoft Virtual Earth.

References

1. Google Maps. http://maps.google.com.
2. MapServer — open source development environment for constructing spatially enabled Internet-Web applications. http://mapserver.gis.umn.edu.
3. MSN Virtual Earth. http://virtualearth.msn.com.
4. MySQL — the world's most popular open source database. http://www.mysql.com.
5. NASA World Wind. http://worldwind.arc.nasa.gov.
6. NIST Net — National Institute of Standards and Technology network emulation pack- age. http://snad.ncsl.nist.gov/itg/nistnet.
7. TopoZone — the Web's topographic map. http://www.topozone.com.
8. MapQuest: Consumer-focused interactive mapping site on the Web. http://www. mapquest.com, 2002.
9. PostgreSQL — the world's most advanced open source database, 2004. http://www.postgresql.org/about/.
10. T. Barclay, J. Gray, and D. Slutz. Microsoft TerraServer: a spatial data warehouse.In *Proceedings of the ACM SIGMOD Conference*, W. Chen, J. Naughton, and P. A. Bernstein, eds., pages 307–318, Dallas, May 2000.
11. A. Harwood and E. Tanin. Hashing spatial content over peer-to-peer networks. In *Australian Telecommunications, Networks and Applications Conference*, pages 1–5, Melbourne, Australia, December 2003.
12. Y. Luo, X. Wang, and Z. Xu. Component-based WebGIS and its spatial cache framework. In *Lecture Notes in Computer Science 3129*, pages 186–196. Springer-Verlag, Berlin, Germany, January 2004.
13. R. C. Nelson and H. Samet. A population analysis for hierarchical data structures. In Proceedings of the ACM SIGMOD Conference, San Francisco, May 1987.
14. H. Samet. *Applications of Spatial Data Structures: Computer Graphics, Image Processing, and GIS.* Addison-Wesley, Reading, MA, 1990.
15. H. Samet. *The Design and Analysis of Spatial Data Structures.* Addison-Wesley, Reading, MA, 1990.
16. H. Samet. *Foundations of Multidimensional and Metric Data Structures.* Morgan-Kaufmann, San Francisco, CA, 2005.
17. H. Samet, H. Alborzi, F. Brabec, C. Esperanca, G. R. Hjaltason, F. Morgan, and E. Tanin. Use of the SAND spatial browser for digital government applications. *Communications of the ACM*, 46(1):63–66, January 2003.

18. E. Stolte, C. von Praun, G. Alonso, and T. Gross. Scientific data repositories: designing for a moving target. In *Proceedings of the ACM SIGMOD Conference*, pages 349–360, San Diego, CA, June 2003.
19. C. Yap, K. Been, and Z. Du. Responsive thinwire visualization: Application to large geographic datasets. In *Proc. 14th Ann. Symp., Electronic Imaging 2002*. IS&T/SPIE, 2002. 19-25 Jan, 2002, San Jose, California.
20. R. Zimmermann, W.-S. Ku, and W.-C. Chu. Efficient query routing in distributed spatial databases. In *Proceedings of the 12th ACM International Workshop on Advances in Geographic Information Systems*, I. F. Cruz and D. Pfoser, eds., pages 176–183, Washington, DC, November 2004.

Chapter 2: Case Study: Geospatial Processing Services for Web-based Hydrological Applications

Laura Díaz, Carlos Granell, Michael Gould

Centre for Interactive Visualization

Department of Information Systems

Universitat Jaume I

Castellón, Spain

Abstract River discharge is an important issue to be monitored because of its significant influence on environmental systems, on human lives for water resource exploitation, and hazards related to floods and landslides. In this context, we have designed and developed a Web-based Geoportal for hydrological applications that integrates geospatial processing services and Web mapping viewers to provide an interactive and user-friendly interface to hydrological modeling experts and scientists. The server side provides hydrological model logic through a library of distributed geospatial processing services that prepares and presents all geospatial data –satellite imagery, cartography, digital elevation models, and sensor measurements– necessary for running the hydrological (river runoff) model .The Geoportal's client side facilitates catalogue service searching for appropriate geospatial data, interacts with the geoprocessing services according to the hydrological model parameters, and displays the results into a Web mapping viewer by using the Google Map API to provide quick feedback to scientists about the status and behaviour of the hydrological model. This chapter provides an overview of the proposed Geoportal by integrating standards both for geospatial processing services and for geospatial data visualization. We emphasize the challenges and problems encountered during implementation regarding the interoperability of different geospatial standards and components.

1 Introduction

Geospatial Web services –Web services that serve and process geospatial information for a wide range of territory-based applications have evolved to become interoperable pieces used to build modular and distributed GIS applications over Internet [18]. They have become key components of Spatial Data Infrastructures (SDI) [15], helping support some the most common requirements of scientific users of information systems such as discovery, access, process and visualization of geospatial datasets.

Scientists traditionally have been major consumers and collectors of huge amounts of data. As in the case of the field of hydrology, they have different needs than geospatial information users from the more cartographic side of the Geographic Information Systems (GIS) field. However they have in common the need to gain remote access (without huge downloads) to large quantities of data, and also to process them remotely using on-line services. Following the e-Science philosophy of connecting scientific research (science) [6], our aim here is to connect scientists to their data, resources, models, and services what is also an important challenge for SDI. Indeed such infrastructures are beginning to facilitate access to distributed, heterogeneous geospatial data through a set of policies, common rules, and standards that together help improve interoperability [17]. Traditional discovery and visualization-based SDI is evolving to a more service-based vision in which geospatial Web services are used not only to access geospatial data, but also to transform them and process them, often in geospatial service chains [1][12].

The recently-published INSPIRE Directive[3] aims to harmonize spatial information across Europe and to improve geospatial data services according to common principles. Our goal here is to help hydrologists –and scientists in general interested in geospatial information– in approaching to INSPIRE philosophy to more efficiently meet their requirements. Hydrologists provide their knowledge and expertise, and SDI researchers play an active role to provide proper geospatial processing services, components and applications to facilitate connecting such hydrologists with their data and models within an interoperable architecture. In this way, the cooperation between data providers and users is fostered as proposed by the [4].

This chapter presents the design and implementation of a concrete Web-based Geoportal for hydrological applications that integrates geospatial processing services and Web mapping viewers like Google Maps to provide an interactive and user-friendly interface to run hydrological models. Our experience in distributing data and processing in the field of hydrology may be extrapolated to other specialized application fields like flooding, forest fires, urban and geological modeling, making then the SDI and the INSPIRE initiatives available to a broader audience.

The remainder of this chapter is structured as follows. Section 2 presents an introduction to hydrological models. Section 3 gives a brief state of the art of the geospatial services and applications for hydrological models. The description of the Geoportal application is the subject of the Section 4 that reviews the system architecture and the underlying open source components and technologies used for

[3] http://inspire.jrc.it/

[4] http://www.gmes.info/

implementation. Section 5 describes the set of geospatial processing services and how are integrated into the Geoportal. Some lessons learnt with respect to user interface issues and the integration and interoperability of geospatial processing services together with the some conclusions are discussed in Section 6.

2 Hydrological Models

River discharge is an important issue to be monitored because of its significant influence on environmental systems, on human lives for water resource exploitation, and hazards related to floods and landslides. In this sense, hydrological models have gained more attention because they provide a physical representation of the hydrological processes occurring in a given basin. The application of these models together with current technologies make possible to monitor and forecast river discharge in a better way. The first task for supporting such monitoring applications is to identify the people who would use the software. Often data providers, who own the data necessary, and scientists, experts in hydrological models, differ in goals and objectives, leading to a lack of collaboration among them.

The AWARE project[5] is a multidisciplinary project carried out by a team of hydrologists, remote sensing specialists, and information system researchers [19]. Its aim is to put together data providers, experts and scientists by developing a user-friendly Web-based prototype that permits not only expert users (hydrologists and other scientists) but also other types of end users and data providers (such as water policy makers, water supply and hydropower companies, irrigation consortia, public authorities) to run concrete hydrological models. This implies usability requirements in our system design leading to an easy-to-use prototype that features intuitive interfaces and wizards assisting non-experts with both the complex tasks of such models and interpreting the results. Since expert users often are more comfortable directly handling and analyzing data, and feel that data provided should be accurately investigated by them, the Geoportal should be flexible enough to serve both experts and novice users. Furthermore, the design of the Geoportal should take into account usability, utility and flexibility requirements.

3 Overview of Available Geospatial Services and Applications for Hydrological Models

Several Web-based hydrological applications are publicly available on the Internet. Most of them are built around a Web mapping service in which several data sets are visualized by applying transparently hydrological model routines. How-

[5] http://www.aware-eu.info/

ever, it is important to highlight some general differences between these Web solutions and our Geoportal. Firstly, our solution allows expert users to interact directly with the underlying hydrological model. A hydrological model normally involves heterogeneous datasets but also several model parameters and variables that must be calibrated. In our application expert users may try several model configurations until the results are acceptable for them. Secondly, in contrast to other applications that use static datasets, expert users should load specific datasets of interest for the area of study. Actually many expert users own the local data necessary to run the model and so the first choice is to allow users to feed the model with the local data they possess. Since a goal of the AWARE project is to be compliant with the INSPIRE initiative, the Geoportal allows other non-expert users to discover and access geospatial data via SDI catalogue services [17]. For instance any user might be interested in searching catalogues for appropriate satellite imagery for the study basin (geographic constraint) and during the snowmelt station (temporal constraint). Finally, another key aspect is that our application is built on distributed geospatial processing services as we detail in the following sections. This aspect meets nicely with the term service-oriented science [6], which refers to scientific research structured as distributed networks of interoperating services. Next we sketch some Web solutions for hydrological applications to provide an overview of current Web-based hydrological applications before describing our Geoportal application in the following sections.

The National Water Information System (NWIS[6]) for the U.S. provides Web-based access to hydrological data to the public and organizations. Basically, NWIS is a data distribution site where users can search and visualize static water data (already embedded in the system), making it impossible to load user data different from those stored in the system. The U.S. Geological Survey (USGS) also offers StreamStat[7], a Web-based tool that allows users to obtain stream flow statistics, drainage-basin characteristics, and other information for user-selected sites. StreamStats users can choose locations of interest from an interactive map and obtain stream flow information for these locations.

A relevant tool is BASINS[8] –Better Assessment Science Integrating Point & Nonpoint Sources from the U.S. Environmental Protection Agency (EPA)–, which is a complete hydrological application for performing watershed-based studied using hydrological models similar to one used in our Geoportal. This application runs on an open source GIS called MapWindow[9], making it more attractive to open source community, yet it is a desktop application that implies that all functionalities and modeling tools are integrated in the application and then performed locally.

---

[6] http://waterdata.usgs.gov/nwis/

[7] http://water.usgs.gov/osw/streamstats/

[8] http://www.epa.gov/waterscience/basins/

[9] http://www.mapwindow.com/

The IJEDI WebCenter for Hydroinformatics [23] is an online application to identify drought-vulnerable regions. The authors propose a combination of data mining techniques to characterize the behaviour of water basins and classify them according to the drought index. Although the goals of IJEDI and our application are slightly different, both deal with multiple kinds of data that have to be integrated and also provide a friendly user interface to be used by non-experts and experts users indistinctly. However, it is important to note that geoprocessing capabilities are not present in the IJEDI application in terms of distributed geospatial services just as our application does.

None of the previous applications execute hydrological models by using distributed geospatial processing services. Certain types of applications demand a distributed approach for multiple reasons such as efficiency and reliability. For example, Web applications for spatial visualization often fully rely on the server to receive the data and visualize them, however some store data on the client side (cache) to handle them locally and improve the response time of data visualization [3]. In our context expert users also try to process large quantities of data remotely using on-line services rather than downloading the required data and processing them locally [8]. Indeed transferring large amounts of data from servers to clients can slow down the whole process, due to network problems or if the server is heavily loaded. Some efforts to define interfaces to access and process multiple kinds of geospatial data remotely are carried out by the Open Geospatial Consortium (OGC). Some basic interfaces (WMS, WFS, etc.) have been already applied to create Web applications [2][4], yet these are shown to be insufficient to suit the specific processing and modeling requirements of hydrological applications. However, the recent OGC Web Processing Service (WPS) specification version 1.0 [21] provides interfaces for interacting with geospatial services by either creating them from scratch or wrapping existing off-line services as Web services. In short, WPS offers three methods to provide the functionality of a certain geospatial processing service by first using the *getCapabilities* method, common in other OGC services, in order to know the available service methods. The WPS defines input and output parameters in a very detailed way by providing a *describeProcess* method. Finally, the *execute* method actually invokes the geospatial processing service with concrete input parameters and returns the results. Here then we propose a Web application that takes advantage of the distributed processing capabilities for performing hydrological models. Similar approaches using WPS services have been taken in [7][11][24], though with some differences regarding our approach as explained in the next section.

This brief overview shows some relevant Web applications for hydrological models pointing out that no Web applications using distributed processing are present in the field of hydrology. The next section details the architecture of our Geoportal application that supports distributed geospatial processing services.

4 System Architecture and Software Components of the Geoportal Application

The software components and the system architecture of the Geoportal application are illustrated in Figure 2.1. The architecture follows a middleware approach composed of three layers. The presentation layer contains the software components used for the user interface (top Figure 2.1). Different servers and distributed geospatial processing services form the middleware layer (centre Figure 2.1) whereas geospatial data and database systems take place in the data layer (bottom Figure 2.1). Here we focus mainly on the components involved in the presentation and middleware layers, describing briefly the data layer. It is important to note that the Geoportal has been built entirely with open source components and technology.

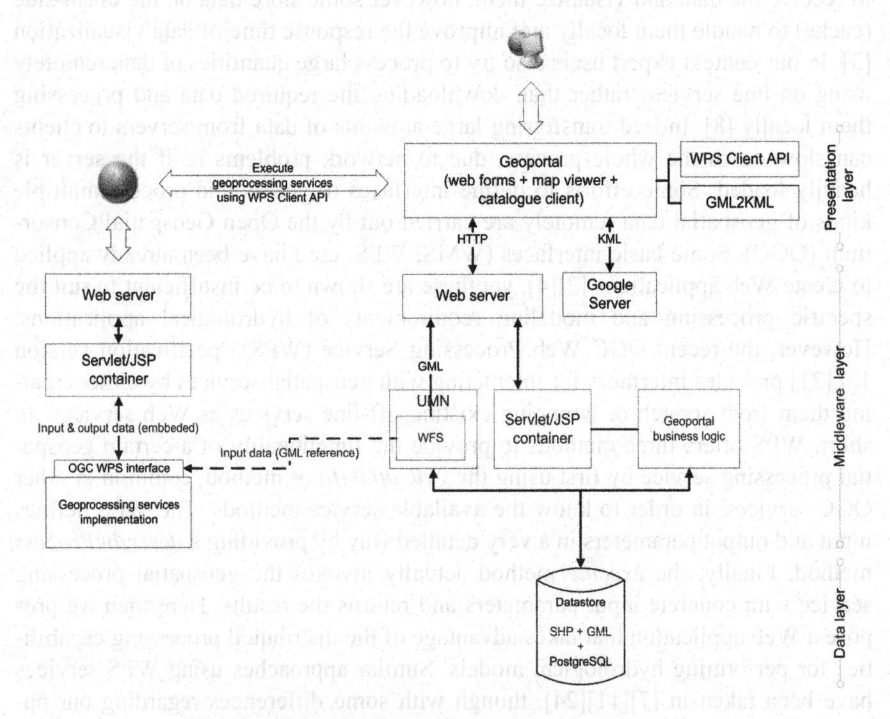

Fig. 2. 1. Geoportal architecture and its components

The presentation layer provides the Geoportal user interface, which permits expert users to select and perform a hydrological model for a concrete basin of study. All users are validated and authenticated when logging in the Geoportal. This feature refers to the possibility to permanently store the current user session in a database. We use for the Geoportal the open source database system PostgreSQL[10] to validate users and to store the result of calibrations successfully completed. A user may perform multiple calibrations for the same basin and decide later which one to choose to forecast the basin discharge (the actual goal of the hydrological model).

The user interface of the Geoportal is composed of multiple Web forms (like wizards) implemented using Apache Struts[11], an open source framework for building Web applications. These Web forms communicate with a set of Java servlets and Java Server Pages (JSP) to offer jointly the Geoportal business logic. The former rely on Apache HTTP Server[12] and the latter on Apache Tomcat[13]. Both servers are integrated using HTTP connectors, which forward internally Apache HTTP Server requests to Apache Tomcat, enabling then the execution of Java servlets through the Apache Server port.

A combination of Java servlets and JSP form the Geoportal business logic in the middleware. They perform a great range of functions such as guiding users through the Web forms to search available data in catalogues, data preparation and collection, data modeling, calibration of model parameters, and interpretation of results. Other common Web-based functionalities such as user authentication are also implemented as Java servlets. If some geospatial processing routine is necessary when a Web form is filled out, the Geoportal is able to invoke the corresponding geospatial processing service through the WPS Client API component. The WPS Client API is a self-developed, key component, because it enables the communication between the presentation and middleware layers, facilitating connection, access and combination of distributed WPS services. It transforms user requests from the presentation layer into OGC WPS requests addressed to a concrete geospatial processing service. Once results are returned, it transforms responses to be properly delivered to the presentation layer (users). In particular, the WPS Client API is developed in a modular way being easily extended to support next versions of the OGC WPS specification. At this moment it supports connectivity to OGC WPS version 0.4 because version 1.0 is, at the time of this writing, still pending approval. Like in the OGC WPS specification, it provides support for the data types specified by OGC such as simple data types and GML (version 2.x). Also, the client API provides a simple caching method for each requested WPS that is invoked for the first time, caching the responses of the invoked *getCapabilities* and *describeProcess* requests. Because a process may be invoked several times, this caching increases processing speed when the same process of a WPS is requested again.

Another key component embedded in the user interface is a Web mapping client or map viewer [16]. In our case we use Google Maps API [10] because it provides a friendly, interactive user interface for novice users. It offers good perform-

---

[10] http://www.postgresql.org/

[11] http://struts.apache.org/

[12] http://httpd.apache.org/

[13] http://tomcat.apache.org/

ance for rendering spatial data, and already incorporates high resolution satellite imagery and other interesting spatial data (for example road network layer in hybrid and map views), very useful to provide context for hydrological applications.

Figures 2.2 and 2.3 show the Geoportal user interface in different steps for the calibration phase of one of the hydrological models. Both figures refer to the Mallero river basin (319 km^2), in the Italian Alps, which is one of the AWARE project test basins. In Figure 2.3, users can check the distribution of precipitation sensors according to the basin boundary. The term *mashup* is currently gaining momentum in describing integrated heterogeneous Web data [9][13] and it is also crucial for our application. Google Maps is a key to our service mashup because it transparently combines both remote data such as the base image and local data such as the basin boundary and the network of precipitation sensors. In addition, users may also click on each red pushpin to get more detailed information such as the location, elevation and name of the corresponding precipitation sensor.

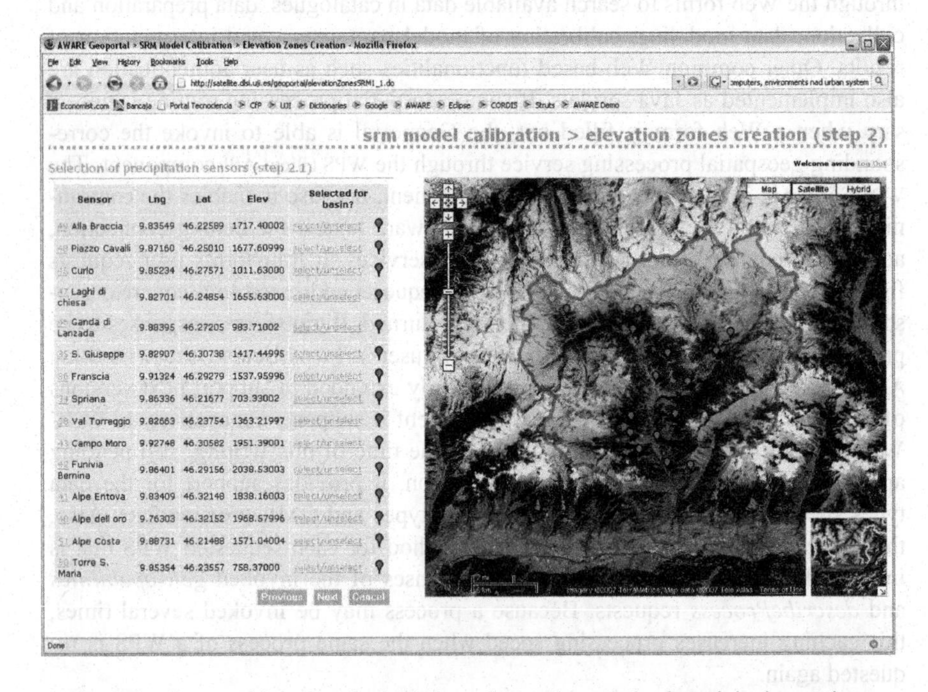

Fig. 2. 2. Geoportal interface for basin boundary and network of precipitation stations

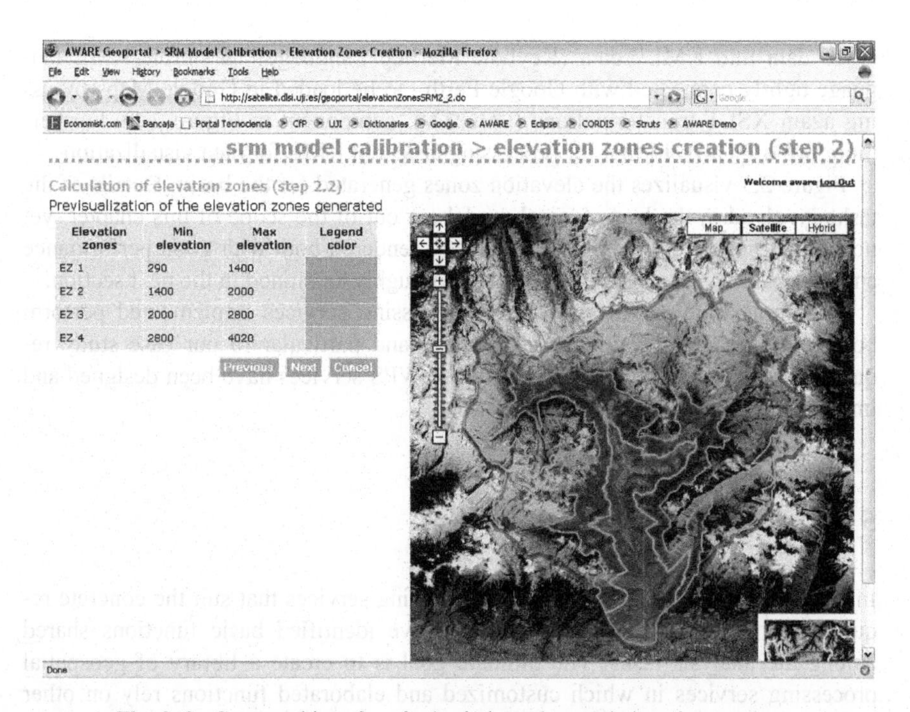

Fig. 2. 3. Geoportal interface for basin boundary and elevation zones

The table of precipitation sensors listed in Figure 2.3 displays a calculated column (fourth) as a result of invoking a WPS service. As mentioned previously, the middleware layer consists of a library of geospatial processing services. At this point, it is necessary to describe how information flows when executing such WPS services. Suppose the example of calculating the elevation given a sensor location (a pair of coordinates) and a DEM (Digital Elevation Model) file for the basin. When the geospatial processing service is required, the application interacts with this service via the WPS Client API that builds appropriate XML-based queries according to the method invoked. Once reported details of the input and output parameters of that process, the WPS Client API has two possibilities (see Figure 2.1) to built the *execute* query: either input data can be embedded in the query itself or can be passed by reference specifying a valid URL to remotely fetch such data. So, sensor location values are embedded in the very XML request while the DEM file is referenced by indicating its URL. The same is applicable to (huge) GML data when are used in a service. In this case, GML data can also come from querying a WFS as occurs in our Geoportal (see Figure 2.1). Both tasks of embedding GML data in WPS *execute* queries and extracting them from WPS *execute* responses are performed by the WPS Client API using XSLT transformations. Finally, when results are forwarded to the presentation layer, the Geoportal transforms the

GML data into KML[14] data (Keyhole Markup Language) –a simple XML language tightly connected with Google Earth– to be loaded in Google Maps by using again XSLT transformations (GML2KML component in Figure 2.1). Furthermore, we use GML format for processing tasks but KML for data visualization.

Figure 2.3 visualizes the elevation zones generated for the basin. Details of the tasks involved in the hydrological model are out of the scope of this chapter, yet notice that huge amounts of KML data are rendered both with good performance and transparency. This example will be thoroughly examined in the next section.

In summary, all of the geospatial processing services implemented perform both basic, general geoprocessing routines and particular to our case study requirements. Next section details how these WPS services have been designed and implemented in our Geoportal application.

5 Geospatial processing services

In order to provide useful geospatial processing services that suit the concrete requirements of hydrological models, we have identified basic functions shared among the analysis tasks. The ultimate goal is to create a library of geospatial processing services in which customized and elaborated functions rely on other much more simple, atomic and well-tested functions. In this way, the reuse of geospatial processing services is fostered because the process of creating new complex geospatial processing services is made possible by mainly reusing already available geospatial processing services from shared libraries [5].

Our design strategy begins by identifying the atomic functions required for the use case. Then we consider a suitable basic geospatial processing service as one which performs a basic function, can be easily tested and is domain-independent enough to be applicable to other contexts. Some examples are geospatial processing services concerned with topological relations such as intersect with, within, crosses, contains, etc., as well as methods for calculating geospatial proximity or distances among geospatial objects and spatial operations like buffer, area and volume. On the other hand, customized geospatial processing services can be defined as those built on basic geospatial processing services to create more elaborated, customized, and domain-dependent. This kind of services normally performs a specific task in a certain domain which cannot be applicable to other contexts. In this sense, they are similar to the concept of opaque or aggregate service chaining defined by OGC as one approach for Web service chaining [1].

Once identified the services, they are grouped into modules with similar functionality. Transforming them into executable WPS is straightforward: each module is a geospatial processing service (WPS) whereas each function either basic or customized is implemented as a process served by the WPS *describeProcess* inter-

[14] http://code.google.com/apis/kml/documentation/

face. Table 1 shows the WPS library with the available processes. Only ElevationZoneCalculation and SnowCoverAreaCalculation are customized processes.

Table 2. 1. List of WPS services in the Geoportal. Bold denotes processes already working in the Geoportal

WPS module	Concrete geospatial processing services
Topology WPS	**Area**, **Intersect**, Union, Contains, Buffer, MaxExtent
ImageProcessing WPS	**Slicing**, SlicingRanges, **Vectorize**, **CoordinateElevation**, **StationsElevation**, **HypsometricElevation**, **ElevationZoneCalculation**, **SnowCoverAreaCalculation**
Chart WPS	**DepletionCurvesPlot**, **DischargePlot**
CoordinatesTransformation WPS	TransCoordGMLPoint, **TransCoordPoint**, TransCoordPoint7P
DataFormat WPS	TransSHPEPSG, TransSHPtoGML

For the implementation of the WPS library we have chosen the OGC WPS specification implementation [21] from the 52° North Open Source Initiative[15], which is an open source platform developed in Java. By using this framework, which provides us with the WPS interface to connect to, we have implemented the algorithms required for the processes listed in Table 1.

To illustrate one basic process we can see how the Area process within the Topology WPS works. First, the Geoportal connects through the WPS Client API to the Topology WPS. Then, when the *describeProcess* interface is requested, a Java object WPSProcess (internal class in the WPS Client API) is instantiated containing information of the process demanded. In this case, this object will specify that the Area process requires a geometry figure like a *Polygon Collection* in GML format. The Geoportal thus sends a *execute* request through the WPS Client API with the basin polygons in GML format, and gets an *execute* response from the Area process containing a real number specifying the basin area. Again, the WPS Client API extracts this value and forwards it to the Geoportal.

Figure 4 illustrates how a customized geospatial processing service is implemented as a chain of basic WPS controlled by the customized process itself. Each basic process is performed as explained previously yet in this case only the final result (step 8) is forward to the presentation layer. In particular, the SnowCoverAreaCalculation process in Figure 2.4 is composed of a sequence of four basic processes: Vectorize, SHP2GML, Intersect, and Area.

Figure 5 shows the ImageProcessing WPS and how it is integrated in the system architecture (see Figure 2.1). This WPS offers processes related to raster image operations like slicing or classifying, where each image cell is classified into categories according to a threshold or a range and a concrete image band. Also it provides a Vectorize process that creates vector data as a set of GML polygons repre-

[15] http://52north.org/

senting the previous classified image. This WPS allows users to extract and process needed information along the model execution without being continuously managing the DEM file, whose size is sometimes too large to work efficiently.

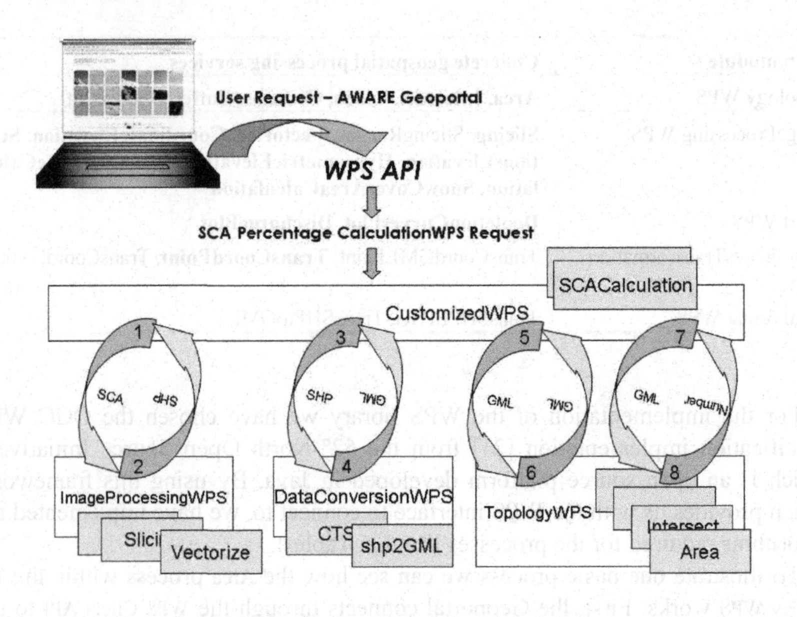

Fig. 2. 4. Snow Coverage Area calculation WPS

Another example is the geospatial processing service for creating the elevation zones (see results in Figure 2.3). To implement this complex task we have designed the customized ElevationZonesCalculation process belonging to ImageProcessing WPS module. Given a DEM file and an elevation range as inputs, this customized process classifies, vectorizes and extracts the polygons in GML format corresponding to the elevation zones. Figure 2.6 shows graphically the steps performs within the ElevationZonesCalculation process. Firstly, scientists search for DEM data for the study basin in available catalogue services. Once DEM references are found, the Geoportal invokes the ImageProcessing WPS service via the WPS Client API. This service uses internally a few open source libraries such as JAI[16] (Java Advanced Imaging API) for the image processing, JTS[17] (Java Topology Suite), GeoTools[18] for the geometric model, edge detector open source software to vectorize the images, and the GML parser integrated in the 52° North WPS implementation to generate and return GML format. We have tested with the ElevationZoneCal-

[16] http://java.sun.com/javase/technologies/desktop/media/

[17] http://www.vividsolutions.com/jts/

[18] http://geotools.codehaus.org/

culation service implementation that interoperability and integration of all of these open source components are possible, though encountering some problems as discussed in the next section.

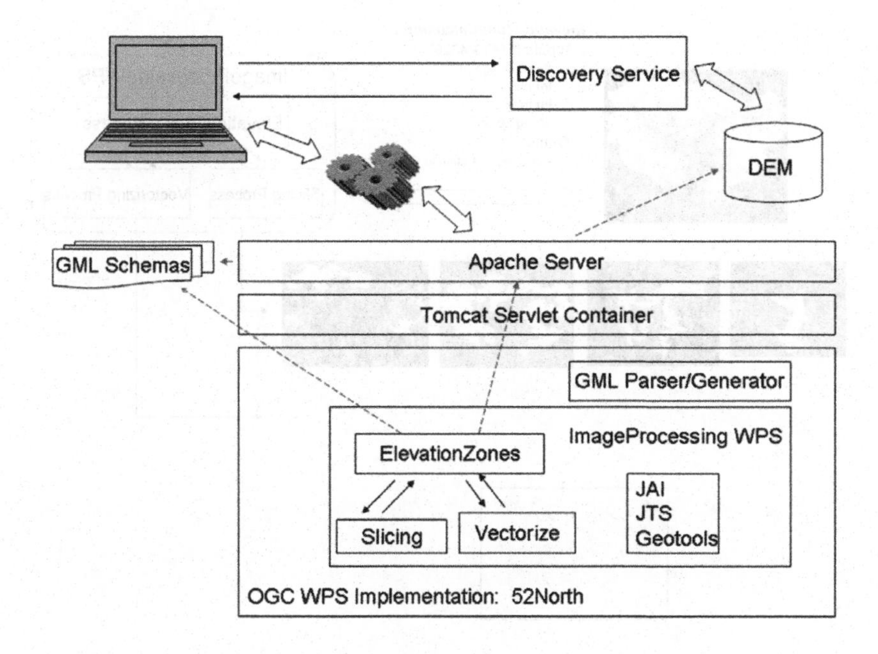

Fig. 2. 5. Image Processing WPS

6 Conclusions and lessons learnt

We have presented a Web-based application that guides expert users in running hydrological models by processing data within a set of distributed geospatial processing services. We have also tested that complex functionality can be processed by connecting to simple, distributed and basic services which can be reusable in other different scenarios. The use of standards is crucial to implement such an application, making it possible in principle the interoperability among all the components involved. In addition the Geoportal has been built on top of an open SDI infrastructure, taking advantage of its benefits and leading to interoperable open software architecture for hydrological applications. We have also shown how it is technically possible that SDI can be used to solve real issues in a more flexible and scalable manner than ad-hoc and stand-alone applications [20]. A pending issue in our work is to provide suitable search mechanisms to find efficiently distributed data and services. In SDI context, these search mechanisms are traditionally catalogue services in which metadata records are essential to describe data and services [22]. However, we find that service metadata need still further research, especially for discovery of WPS services.

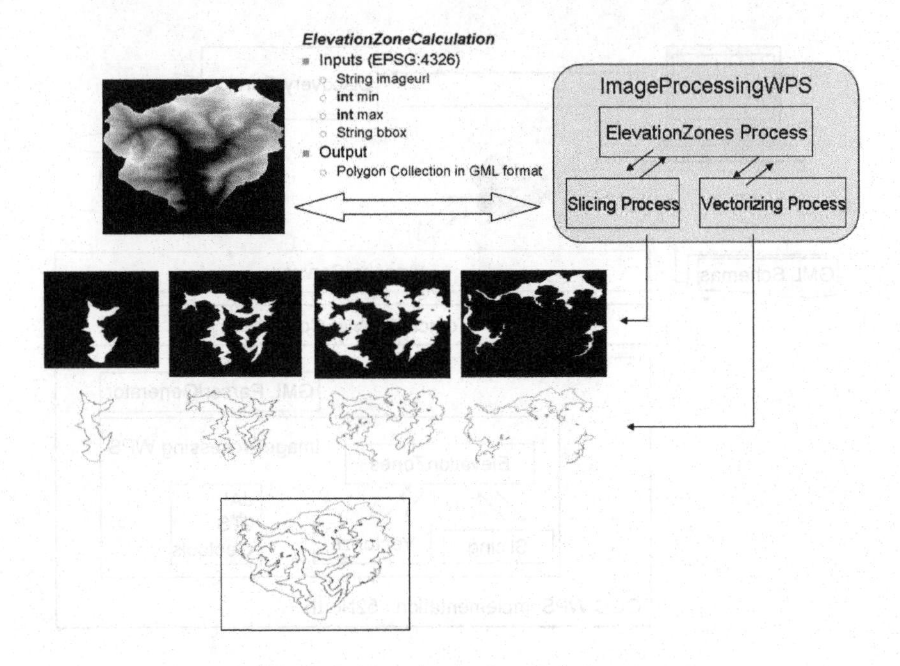

Fig. 2. 6. Elevation zones calculation

Although theoretically the use of standards should be sufficient to achieve interoperability, we found that each vendor implementation differ from each other. Some decisions taken with regards to specific vendor implementations have had a great influence in the target application, making it difficult to reach interoperability at programmatically level (in practice). Some lessons to keep in mind about data integration would be that simple, structured data (e.g. KML) is easier to manage and process than powerful but complex data (e.g. GML) which is more sensitive to failure when processing information. It is assumed that GML has great advantages as a language to integrate disparate formats and to serve data through services [14]. For instance GML documents permit representing complex spatial models by nesting geographic features in a XML way. Yet we have experienced more difficulties handling GML data rather than KML data. In some cases where data complexity is not a system requirement, KML may be a valid, efficient mechanism for exchanging and visualization geospatial data.

Extracting information from huge GML documents has been revealed as an important issue. Processing a GML document may become sometimes challenging due to low performance when working with huge data sets. Beyond text overhead and verbosity inherent in GML (and XML in general), this limitation is caused by the latency produced in data transfers between servers and clients. In our particular case, one way to overcome partly this limitation has been to work with referenced data rather than the GML data themselves. When different basic WPS services and customized WPS (chains) are called, we make reference to the GML data which are only transferred when they are really needed for processing or for visualization purposes.

Other important issue when working with GML is connecting GML data created by different vendor's services. Schemas created by different OGC services implementations are not completely valid, so they cannot be validated by most of the XML readers and then they cannot be parsed by most of the open source GML parsers [14]. Each application creates their own schemas that work normally only within its particular context. We performed extensive research on GML and GML schema, ending with a valid GML to produce basins and elevation zones. The flexibility and extensibility of GML can be then seen as a weak point at practice level when talking about interoperability. It has been a difficult task to generate GML data according to the needed schemas that were successfully parsed by all the GML parsers and generators used in the Geoportal. For example, it was relatively easy to generate a simple subset of GML that was accepted by components involved in the Geoportal like 52° North WPS Implementation and the GeoTools library. The compatibility problems arose when we tried to use standard and complete GML files together with other GML produced by other software components, in order to be transparently used by our WPS services. It was not possible, to our best knowledge, to find a general, valid schema able to be used for every GML files generated by the different software tools and technologies used in the Geoportal.

Integrating different open source libraries implies sometimes a high development complexity, because quite often on-going projects are not very stable and one becomes part of the testing team, facing development bugs which have to be solved. This is even more accentuated when the project belongs to a recent research technology, which is the case of our Geoportal implementation. The OGC WPS specification seems at the moment to be sufficiently mature to be implemented as we found using the 52° North WPS Implementation. However it is still in experimental phase.

Finally, we encountered problems in the implementation of the Geoportal user interface due to multiple projection systems. Google Map viewer uses a common geographic projection that refers to WGS84 (EPSG code 4326) as a pair of coordinates (longitude, latitude), yet geospatial data are given and processed, normally in hydrological applications, in distinct projection systems. For this reason, coordinate transformation services are necessary in the Geoportal, however, it will increase the response time to the user.

A first observation derived from our experience in this project, which coincides with conclusions in previously referenced work, is that the approach based on distributed geoprocessing services leads to a collection of reusable geospatial processing services, available for other users in the case that they are well-documented and registered in open catalogues. This is possible in principle because WPS geospatial processing services do not work with pre-established datasets but rather they preserve a loosely-coupled relationship between data and processing capabilities (algorithms), making it possible to chain them to other geospatial Web services such as WMS and WCS. One of the problems being partially addressed in the Geoportal application is when a geospatial processing service exchanges and processes large amount of data, which still needs further research for the geospatial community.

Acknowledgments: This work has been partially supported by the AWARE project SST4-2004-012257 co-funded by the EU and the GMES initiative. Institut Cartogràfic de Catalunya (ICC) assisted in the design of the Geoportal.

References

1. Alameh N (2003) Chaining Geographic Information Web Services. IEEE Internet Computing 7(5):22-29
2. Anderson G, Moreno-Sanchez R (2003) Building Web-Based Spatial Information Solutions around Open Specifications and Open Source Software. Transactions in GIS 7 (4): 447–466
3. Brabec F, Samet H (2007) Client-Based Spatial Browsing on the World Wide Web. IEEE Internet Computing 11(1): 52-59
4. Caldeweyher D, Zhang J, Pham B (2006) OpenCIS-Open Source GIS-based Web community information system. International Journal of Geographical Information Science 20: 885-898
5. Díaz L, Costa S, Granell C, Gould M (2007) Migrating geoprocessing routines to Web services for water resource management applications. In Proceedings of 10th AGILE Conference on Geographic Information Science (AGILE 2007), Aalborg (Denmark)
6. Foster I (2005) Service-Oriented Science. Science 308: 814-017
7. Friis-Christensen A, Bernard L, Kanellopoulos I, Nogueras-Iso J, Peedell S, Schade S, Thorne C (2006) Building service oriented applications on top of a spatial data infrastructure – a forest fire assessment example. In Proceedings of 9th AGILE Conference on Geographic Information Science (AGILE 2006), Visegrad (Hungary)
8. Granell C, Díaz L, Gould M (2007) Managing Earth Observation data with distributed geoprocessing services. In Proceedings of the International Geoscience and Remote Sensing Symposium (IGARSS 2007), Barcelona (Spain)
9. Jhingran A (2006) Enterprise information mashups: integrating information, *simply*. In Proceedings of the 32nd international Conference on Very Large Data Bases. VLDB Endowment, 3-4.
10. Jones MT (2007) Google's Geospatial Organizing Principle. IEEE Computer Graphics and Applications 27(4): 8-13
11. Kiehle C (2006) Business logic for geoprocessing of distributed geodata. Computers & Geosciences 32: 1746-1757
12. Lemmens R, Wytzisk A, de By R, Granell C, Gould M, van Oosterom P (2006) Integrating Semantic and Syntactic Description to Chain Geographic Services. IEEE Internet Computing 10(5): 42-52
13. Liu X, Hui Y; Sun W; Liang H (2007) Towards Service Composition Based on Mashup. In Proceedings of 2007 IEEE Congress on Services, 332-339
14. Lu C-T, Dos Santos R, Sripada L, Kou Y (2007) Advances in GML for Geospatial Applications. GeoInformatica 11(1): 131-157

15. Masser I (2005) GIS Worlds; Creating Spatial Data Infrastructures. ESRI Press, Redlands, CA
16. Mitchell T (2005) Web Mapping Illustrated. O'Reilly Media, Sebastopol, CA
17. Nogueras-Iso J, Zarazaga-Soria J, Muro-Medrano P (2005) Geographic Information Metadata for Spatial Data Infrastructures – Resources, Interoperability and Information Retrieval. Springer, Berlin
18. Peng Z-R, Tsou MH (2003) Internet GIS: Distributed Geographic Information Services for the Internet and Wireless Networks. Wiley, Hoboken, NJ
19. Rampini A, de Michele A, Lehning M, Blöschl G, Brilly M, Llados A, Sapio F, Gould M (2006) AWARE: A tool for monitoring and forecasting Available Water Resource in mountain environment. Geophysical Research Abstracts 8(10780)
20. Scholten M, Klamma R, Kiehle C (2006) Evaluating Performance in Spatial Data Infrastructures for Geoprocessing. IEEE Internet Computing 10(5): 34-40
21. Schudt P (ed) (2007) OpenGIS Web Processing Service Version 1.0.0, Open Geospatial Consortium. Available at http://www.opengeospatial.org/standards/wps
22. Smits PC, Friis-Christensen A (2007) Resource Discovery in a European Spatial Data Infrastructure. IEEE Transactions on Knowledge and Data Engineering 19 (1): 85-95
23. Soh L-K, Zhang J, Samal A (2006) A Task-Based Approach to User Interface Design for a Web-Based Hydrologic Information Systems. Transactions in GIS 10 (3): 417–449
24. Yuan Y., Cheng Q. (2007) Integrating Web-GIS and Hydrological Model: a Case Study with Google Maps and IHACRES in the Oak Ridges Moraine area, Southern Ontario, Canada. In Proceedings of the International Geoscience and Remote Sensing Symposium (IGARSS 2007). Barcelona (Spain), July 2007 (in press).

Chapter 3: An Application Framework for Rapid Development for Web-based GIS: GinisWeb

Aleksandar Milosavljević

Faculty of Electronic Engineering

Aleksandra Medvedeva 14, 18000 Niš, Serbia

alexm@elfak.ni.ac.yu

Slobodanka Đorđević-Kajan and Leonid Stoimenov

Department of Computer Science
Faculty of Electronic Engineering

University of Niš

Abstract This chapter presents *GinisWeb*, an application framework designed to enable rapid development of standards based Web GIS. A Web GIS, as a whole, is considered through a client/server model. Web GIS application that enables access to shared geodata is in the role of a client. Geodata capturing, storing, maintaining and sharing through standardized Web interfaces is the responsibility of a Web GIS server side, named *Web-enabled GIS*. Since both client and server side operate on the same set of geodata, the design of the framework relied on a definition of a unique *GinisWeb geoinformation model*. Rapid development is based on an explicit description of application domain that can be used by the framework. To cope with these descriptions, design of the framework included specification of an XML language named *Ginis Application Definition Language* (GADL). A development of a new Web GIS application, supporting desktop GIS application, and Web GIS services (as a part of Web-enabled GIS) using *GinisWeb* framework, rely on the use of configurable application components produced by the framework and GADL encoded descriptions of the concrete application domain. Finally, to evaluate and present results obtained by this work, this chapter includes a case study concerning implementation of a Web GIS for an electric power supply company.

1 Introduction

A geographic information system or geoinformation system (GIS) is a special type of computer-based information system, tailored to store, process, and manipulate geospatial data [20]. The ability of GIS to handle and process both location and attribute data, distinguishes GIS from other information systems. It also establishes GIS as a technology that is important for a wide variety of applications [2]. Traditionally, geographic information systems were built as monolithic and platform-dependent applications [19], but, with the development of the Internet and the World Wide Web, GIS has evolved and adapted to this new environment [14]. 2001). The "Web GIS" became a synonym for Web information systems that provide the functions of geographic information systems on the Web through HTTP and HTML [13].

GIS applications built for the Web environment usually provide functions for map displaying and navigation, as well as functions for querying geodata using both spatial and non-spatial criteria. It is obvious that these functions can cover only GIS functionality of displaying, analyses and integration of geospatial data [15]. For capturing, storing and manipulating geospatial data more traditional application environments are still needed.

A Web GIS, considered as a whole, relies on a client/server model, where client (*Web GIS application*) provides access to geodata, while one or more servers provide their sharing. The sharing of geospatial data for Web presentation purposes can be done by the extension of existing traditional GISs with a set of "Web interfaces". This kind of extension forms a Web GIS server side that we named *Web-enabled GIS.* In order to build a model of a Web GIS that is open for connecting to a variety of different geodata sources, commonly acceptable standards for implementation of these Web interfaces are needed [9]. Currently, Open Geospatial Consortium[19], as an international industrial consortium with an aim of developing publicly available standards in the field of GIS, has several implementation specifications that standardize this field. Our functional needs for Web interfaces were met by two of these specifications. The first is *Web Map Service (WMS) Implementation Specification* [7] that describes a Web service interface for custom maps retrieval, while the second is *Web Feature Service (WFS) Implementation Specification* [16] which describes a Web service interface for querying and retrieval of geospatial entities using both spatial and non-spatial criteria.

To enable rapid development of a class of Web GISs with the same functionality over different geodata sources we relied on the development of an application framework named *GinisWeb*. Framework is a generic term for a powerful object-oriented reuse technique that typically emphasizes the reuse of design patterns and architectures [6]. There are two common definitions of an application framework [5]. The first defines an application framework as a reusable design of the entire,

---

[19] Official Web page: http://www.opengeospatial.org

or part of a system represented by a set of abstract classes and the way their instances interact, while the second states that an application framework is the skeleton of an application that can be customized by an application developer. These definitions are complementary, not conflicting, since the former describes framework from the design perspective, whereas the latter describes it from the functional viewpoint. The primary benefits of object-oriented application frameworks stem from the modularity, reusability, extensibility, and inversion of control they provide to developers [5].

Another framework based approach for rapid development of Web GIS applications is presented in [18]. The presented WebGD framework relies on ArcIMS for providing interactive map images and on ArcSDE for managing spatial data, unlike ours which can be utilized for the development of both client and server side of Web GIS, the presented WebGD framework. WebGD includes a set of ASP.NET custom server controls that provide user interface elements such as a map, a layer list, and a toolbar. Applications developed with WebGD allow users to insert, update, delete, and query data with a map interface displayed by Web browsers.

GinisWeb framework described in this chapter produces configurable application components required for the implementation of the overall Web GIS [10]. For their configuration we need explicit descriptions of a concrete application domain in terms of geodata types and sources. To cope with these descriptions the framework design included specification of an XML language named GADL.

Similar approach for applying XML technologies to increase the efficacy in the GIS application development is presented in [8]. The authors applied XML technologies to improve the robustness of geological and geophysical applications as well as to increase the efficacy of the application development process. Unlike our approach that is based on a framework capable of "understanding" GADL XML descriptions, their approach relies on an XML to C++ binder to automatically generate C++ code for data containers, as well as on parsing, validation, and data object serialization. Data describing, like in our approach, is done using XML schema. The major benefit from our, framework based approach is that the development of a new Web GIS (with standard functionalities) requires only changes in GADL descriptions leaving application components and code intact.

The chapter is structured as follows: in the second part, we describe architecture of an overall Web GIS system in short, identify subsystems, analyze use-cases, specify non-functional requirements, functions and structure of a Web GIS application and Web-enabled GIS, and introduce a developmental model of Web GIS using *GinisWeb* framework. The third part of this chapter is dedicated to *GinisWeb* model of a geoinformation system that is the base for implementation of all components of the suggested overall Web GIS. The fourth part gives introduction to the structure and basic elements of an XML language GADL, while the fifth part presents a case study concerning the implementation of Web GIS for a power supply company. Finally, in the conclusion, the achieved results are summarized.

2 Architecture of overall Web GIS

To identify subsystems of overall Web GIS we started from a general assumption that geoinformation related to some area cannot be captured, stored and maintained in a single organizational unit GIS. Some of this information can have mutual and public importance, so it should be **shared** and **accessible** over the Web.

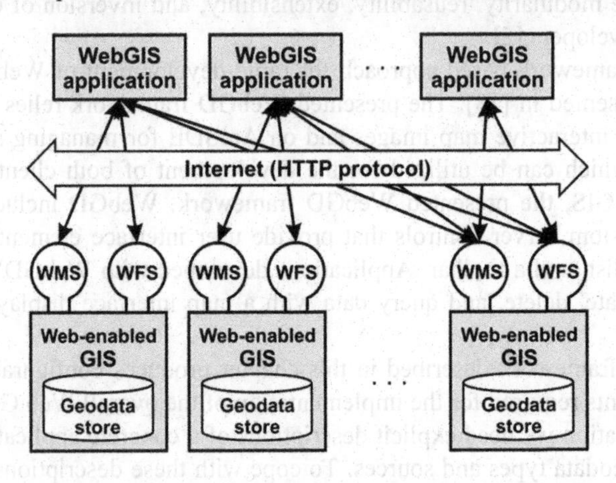

Fig. 3. 1. Schematic view of proposed Web GIS architecture.

A subsystem that enables access to geoinformation over the Web is *Web GIS application*, while the sharing of geoinformation is done by extending traditional GISs with WMS and WFS Web interfaces (*Web-enabled GIS*). The interrelation among components that correspond to subsystems is illustrated in Fig. 3.1. The reliance on standards for implementation of interfaces of a Web-enabled GIS, allows Web GIS applications to provide content from several GIS nodes. This also implies that one GIS node can serve several Web GIS applications. The term *GIS node* is used to signify an instance of Web-enabled GIS subsystem.

2.1 Use-case model of the overall system

The proposed Web GIS is used by at least three classes of users: *Authorized GIS users*, *Web users* and *Web GIS administrators*. UML diagram showing their association with the use-cases is given in Fig. 3.2. Authorized personnel have access to all functions of their local desktop GIS applications which include: map viewing, feature inserting, editing, deleting and querying. Web users are the most general class with the least functionality at their disposal. In order to view maps and query features, they access the system through Web GIS application. Finally, the administrator is responsible for defining contents that will be available to Web users. For those purposes, administrator must define the layer hierarchy and characteristics through which geoinformation will be presented, and also assign adequate WMS and/or WFS services to the layers.

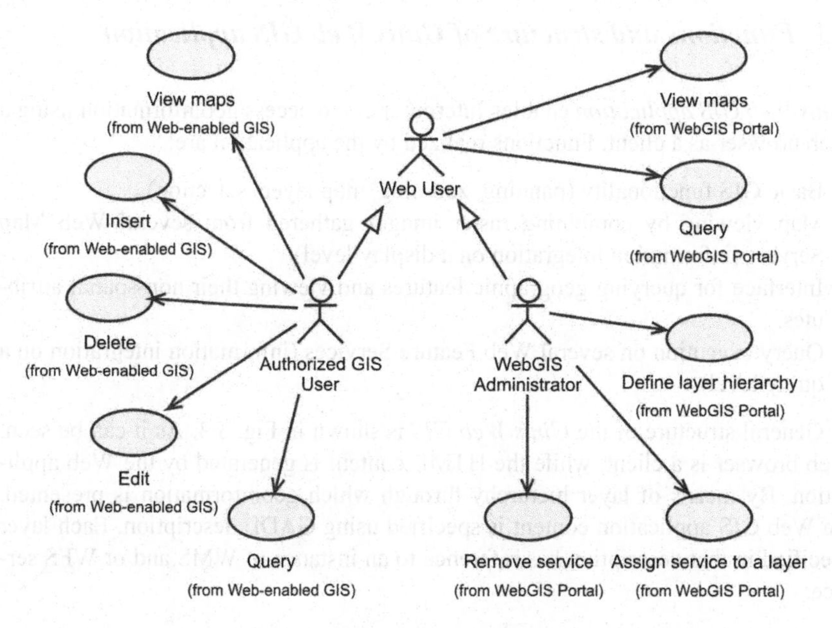

Fig. 3. 2. UML diagram that represents the use-case model of the overall system.

2.2 Non-functional requirements

The most important non-functional requirements for the system that is being proposed are *code and component reusability*, *extensibility* and *openness*.

Reusability is achieved through the designed *GinisWeb* application framework. The framework, through component reusability, and using XML description, should enable development of a class of Web GIS applications and Web-enabled GISs with the same functionality over different domains and data sources. Using it on a code level, it also enables the building of more specialized GIS applications that can easily be ported to the Web.

Extensibility of the system means that a variety of WMS/WFS enabled GISs can be connected to a Web GIS application. Assigning this geoservices to a Web GIS application is done using XML descriptions, too.

A demand for openness imposes the use of interfaces and exchange of data by commonly accepted standards. This requirement is met using OGC standards (WMS and WFS) for Web services interfaces and data interchange. Although

GinisWeb framework offers full support for building an overall Web GIS, it also allows the use of third parity OGC WMS and WFS compliant servers.

2.3 Functions and structure of Ginis Web GIS application

Ginis Web GIS application enables Internet users to access geoinformation using a Web browser as a client. Functions realized by the application are:

- Basic GIS functionality (panning, zooming, map layers selection).
- Map viewing by combining raster images gathered from several Web Map Servers (information integration on a display level).
- Interface for querying geographic features and viewing their non-spatial attributes.
- Query execution on several Web Feature Services (information integration on a query level).

General structure of the *Ginis Web GIS* is shown in Fig. 3.3. As it can be seen, Web browser is a client, while the HTML content is generated by the Web application. By means of layer hierarchy through which geoinformation is presented, the Web GIS application content is specified using GADL description. Each layer specified in that description has reference to an instance of WMS and/or WFS service.

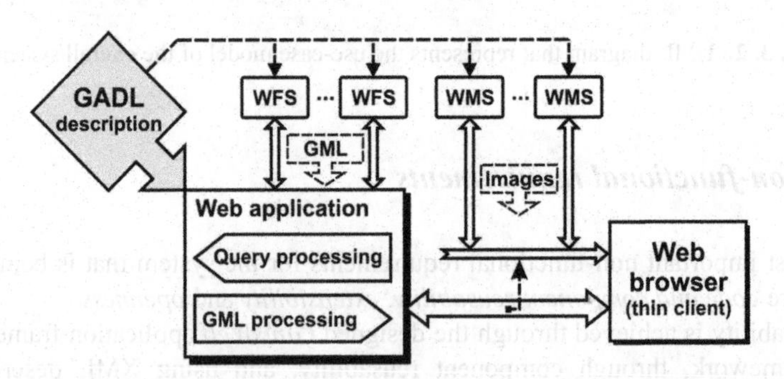

Fig. 3. 3. Structure of Web GIS application built using the GinisWeb framework.

Composing a map using georeferenced raster images gathered from several WMSs is done on a Web browser, while the Web application is in charge of creating a HTML page with corresponding image requests toward map servers. The implementation of feature querying met a more complicated interface of WFS, so the Web application has a more significant role. This implies generation of HTML pages with user interface for specifying search criteria, accepting queries defined in that manner, their processing and the composition of valid WFS requests. Since WFS response is, in general, encoded in GML, the Web application must process the received GML documents and then generate an appropriate HTML page with query results.

2.4 Structure of Web-enabled GIS node

As it was already stated, a GIS node represents an instance of a Web-enabled GIS. Based on open standards, the suggested architecture does not impose any concrete structure and implementation of a GIS node, as long as OGC WMS and WFS are used as Web interfaces. Nevertheless, one of the main goals considering research presented in this chapter is definition and implementation of a framework, flexible enough to allow building of GIS nodes, too.

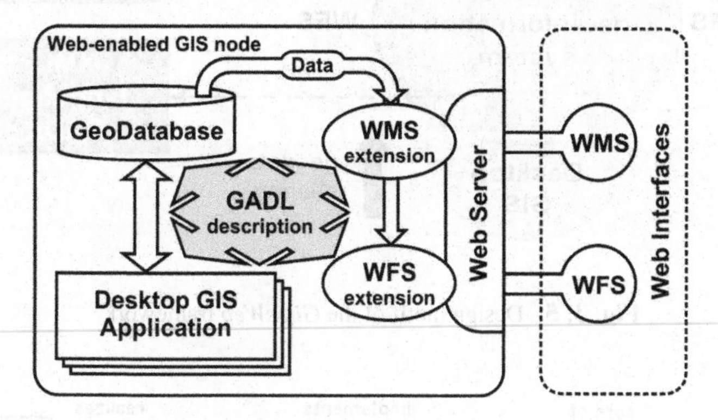

Fig. 3. 4. Structure of Web-enabled GIS node built using the *GinisWeb* framework

Our suggestion for a structure of Web-enabled GIS relies on using standard components such as a geodatabase for storing geodata and a desktop GIS application for geodata capturing and manipulation. Web interfaces are built using specialized Web server extended with WMS and WFS service implementations. For automatic configuration of these reusable components explicit descriptions of a GIS application domain are used. The encoding of these descriptions is done by using XML language GADL in a way similar to Web GIS application. The structure of a Web-enabled GIS node that can be built using the *GinisWeb* framework is shown in Fig. 3.4.

2.5 Development model of the *GinisWeb* framework

The design of the *GinisWeb* framework relies on an assumption that Web GIS application, desktop GIS application, and services that represent Web interfaces to geoinformation are just different interfaces against the unique structure of geoinformation and geoprocessing methods. That unique structure is recognized and represented in ***GinisWeb* model of geoinformation system** (Fig. 3.5).

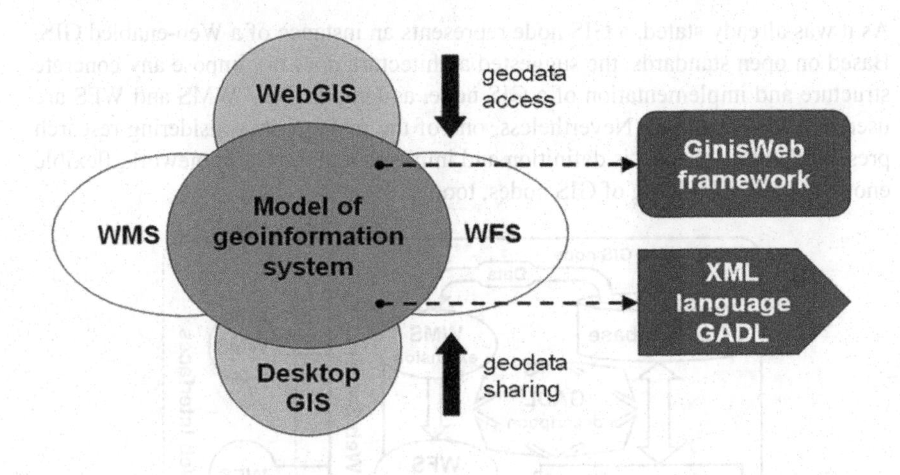

Fig. 3. 5. Design motif of the *GinisWeb* framework

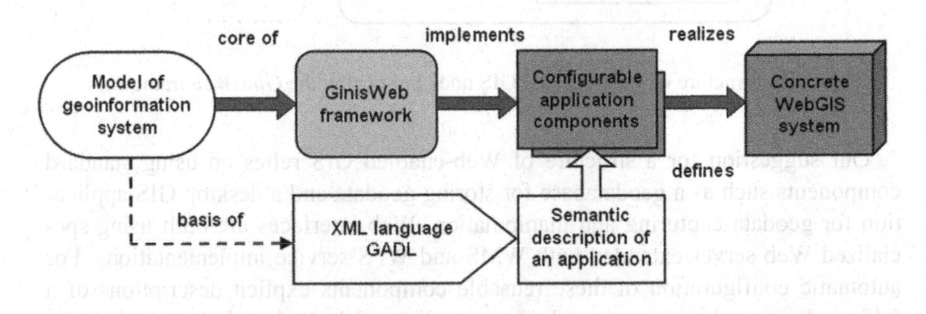

Fig. 3. 6. Development model of a Web GIS based on the *GinisWeb* framework

Elements and structure of geoinformation system encapsulated in *GinisWeb* model represent the core for implementation of *GinisWeb* framework and the basis for specification of the XML language GADL (Fig. 3.5). Configurable application components produced by the first and configured by the latter, lead to a concrete overall Web GIS (illustrated in Fig. 3.6).

Connections and communication between components that construct Web GIS are illustrated using UML collaboration diagrams in Fig. 3.7. The diagram objects WebApp, WMS, WFS, and WinApp correspond to application components produced by the framework. These objects are related to *GinisWeb* framework interface classes that will be described in the next section. GADL objects represent XML documents that describe Web GIS application and Web-enabled GIS. All four application components have appropriate GADL description. Finally, GeoDB and Web browser objects represent geodata store and Web GIS client, respectively. The first scenario (Fig. 3.7a) shows the flow of communication between components in performing map displaying task, while the second (Fig. 3.7b) depicts query execution process. It is obvious that execution of a query is a more complicated task because WebApp is responsible for generating query interface, translating query into a WFS request, processing the retrieved GML, and generating an HTML page displaying results. In a map displaying task WebApp has only the responsibility to generate HTML code with adequate WMS request(s) as image source(s), so that Web browser can directly compose a map using image(s) retrieved from WMS(s).

3 *GinisWeb* model of a geoinformation system

GinisWeb model of a geoinformation system defines organization of geodata and geoprocessing methods used in *GinisWeb* GIS applications. The model is unique and it is used as a core for implementation of *GinisWeb* framework that produces all GIS application components needed by the suggested overall Web GIS (see Fig. 3.6).

The definition of the *GinisWeb* model of a geoinformation system was based on OGC reference model [1] and qualified with some elements stated through WMS [7], WFS [17], GML [3], and Filter Encoding [17] specifications. GML is OGC standard for XML based encoding of data about geographic features. Geographic feature is described with its spatial (geometries) and non-spatial attributes. WFS, as another OGC standard, use GML for encoding of query results, while for encoding of query request it uses OGC Filter Encoding Specification. When WFS is requested to describe some feature type it uses XML Schema Definition Language (W3C standard) to encode that description. While WFS operates on features, i.e. feature types, WMS operates on layers that compose a geographic map. That is why our *GinisWeb* geoinformation model has class Layer as a basic data organization unit, while features and feature types are introduced through FeatureLayer specialization. Feature layers can be both queries using WFS and displayed using WMS interface, while other types of layers (general class Coverage) can only be displayed using WMS interface. The model is presented through UML class diagram shown in Fig. 3.8.

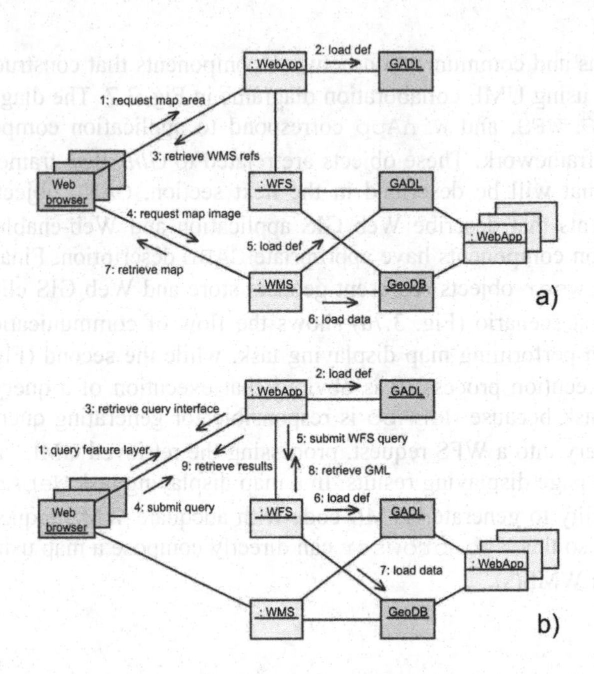

Fig. 3. 7. UML collaboration diagrams that illustrate connections and communication between components in a map retrieval scenario (a); and a query execution scenario (b).

`WinApp`, `WMS`, `WFS`, and `WebApp` classes represent interfaces for the implementation of configurable application components of an overall Web GIS. These classes are derived from the `GisApp` class that encapsulates basic functionality of a GIS application. The `GisApp` class represents specialization of a class `Layer`, the central class of this model responsible for defining a general map layer type. This necessarily means that an application inherits all the functions defined in a layer context. The detailed UML diagram of a class `GisApp` is shown in Fig. 3.9. The method `Login` enables logging in and initialization of an application. Upon successful logging, username and password are stored for further use in an instance of the `UserInfo` class. Attribute `bbox`, represented as an association with `BBox` class, holds boundaries of an area that is currently being displayed within a map. Association with a class `Geometry` through `selectedGeom` attribute indicates currently selected geometry of some feature. A reference to a currently selected layer is contained within attribute `activeLayer`. If an active layer is a feature layer, then attribute `activeGeom` refers to a definition of geometry that is currently active for insertion. The other methods declared in a `GisApp` class are inherited from a class `Layer` with definitions that correspond to GIS application needs.

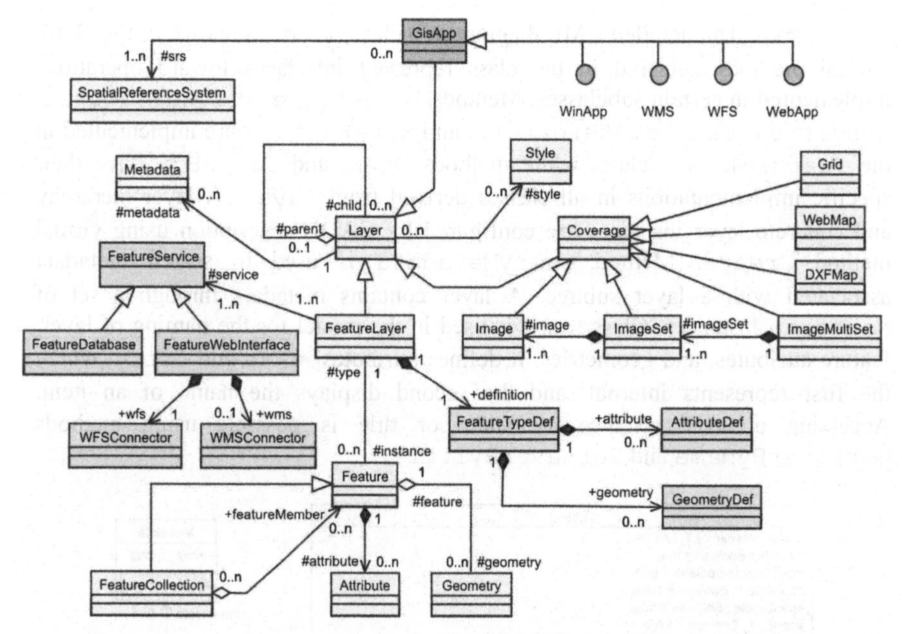

Fig. 3. 8. UML class diagram that represent the *GinisWeb* model of a geoinformation system

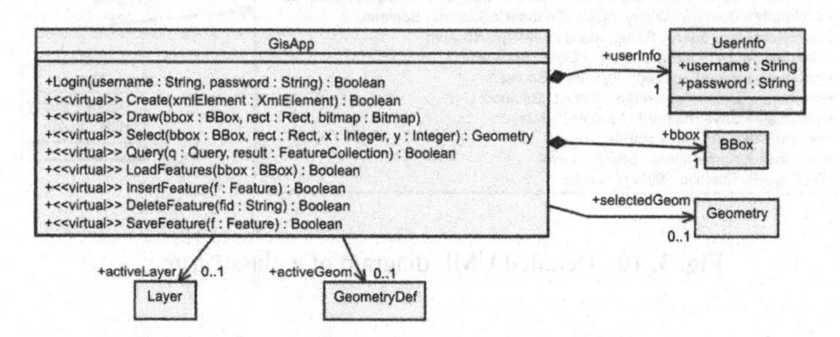

Fig. 3. 9. Detailed UML diagram of a class `GisApp`

The presented model is characterized by the organization of geoinformation into hierarchy of layers (see class `Layer` in Fig. 3.8). Hierarchical organization of layers enables the aggregation and classification of features according to certain criteria. Functions that a layer transmits to its sublayers are display and search. This means that if a layer is included in a display of a map, or a query is submitted on it, all of its sublayers will be shown, i.e., searched as well.

The role of a basic layer class (`Layer`) is to define all function interfaces, implement hierarchical organization, and maintain metadata. To represent different types of geoinformation we use the following `Layer` derived classes: `FeatureLayer`, `Coverage`, `Image`, `ImageSet`, `ImageMultiSet`, `Grid`, `WebMap`, and DXFMap. The detailed UML diagram of a class Layer is shown in Fig. 3.10. Virtual methods declared in the class represent interfaces toward operations implemented in certain subclasses. Methods Select, Query, LoadFeatures, InsertFeature, DeleteFeature, and SaveFeature are implemented in the FeatureLayer class, while methods Draw, and CalcBBox have their specific implementations in all classes derived from Layer. A layer hierarchy and concrete layer instances are configured by GADL description using virtual method Create. Method QueryMetadata is used to search metadata associated with a layer subtree. A layer contains metadata through a set of Metadata instances. Class Name is used in the model for the naming of layer, feature attributes, and geometries. It defines attributes' name and title, where the first represents internal, and the second displays the name of an item. Accessing a layer instance by name or title is possible using methods GetLayerByName and GetLayerByTitle.

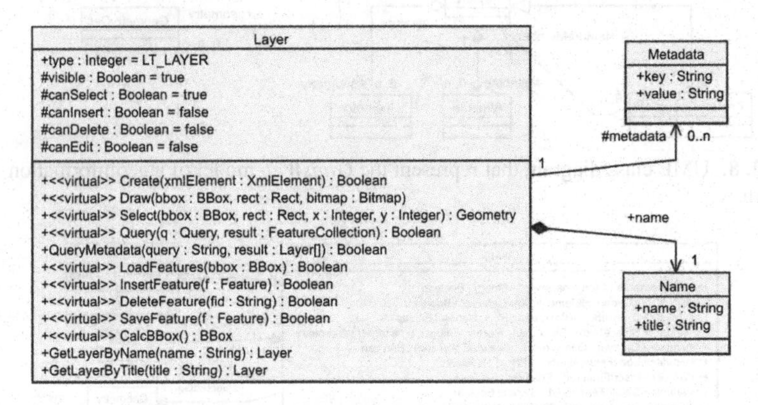

Fig. 3. 10. Detailed UML diagram of a class Layer

Class FeatureLayer defines the layer that holds a collection of features of a certain type (see Fig. 3.11). It contains definition of a feature type (class FeatureTypeDef) through a specification of a set of non-spatial attributes (class AttributeDef) and spatial attributes, i.e. geometries (class GeometryDef) contained by features of that type. A definition of geometry is specified through attributes of a GeometryDef class. Creating the corresponding Geometry object by using an adequate segment of GADL description is done within Create method. A definition of feature non-spatial attributes is more complex and requires additional specifications of value type, default value, and optional value constraints. A set of concrete feature instances of a type are represented by Feature class that holds geodata for the layer through Attribute and Geometry classes.

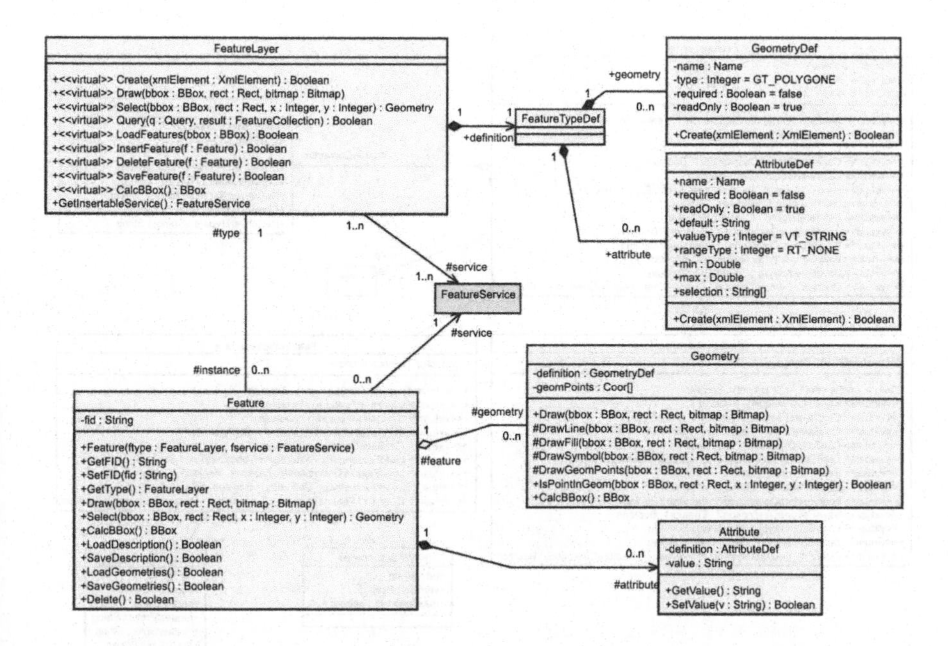

Fig. 3. 11. Detailed UML diagram of classes related to `FeatureLayer`

A set of services related to a `FeatureLayer` is represented by an abstract `FeatureService` class (see Fig. 3.12). Concept of feature service is used to provide and optionally enable storing of features. Methods declared in `FeatureService` enable the following set of activities:

- loading of geographic features of a certain type (method `LoadFeatures`),
- execution of queries on features of a certain type (method `Query`),
- inserting a new feature (method `InsertFeature`),
- deleting a feature (method `DeleteFeature`),
- loading and saving feature attributes (methods `LoadFeatureDescription` and `SaveFeatureDescription`),
- loading and saving feature geometries (methods `LoadGeometry` and `SaveGeometry`).

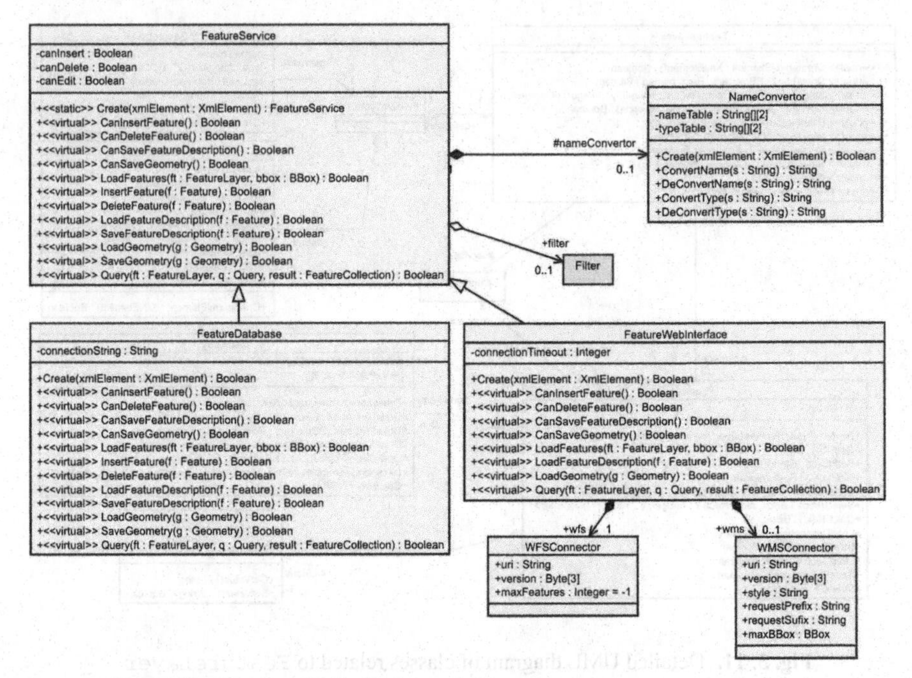

Fig. 3. 12. Detailed UML diagram of class related to `FeatureService`

Class `Coverage`, together with its specializations, defines a type of a layer containing geodata that cover some area (see Fig. 3.13). Classes `Image`, `ImageSet`, and `ImageMultiSet` are used for the display of georeferenced raster images. Methodology for creating such images from scanned paper maps is described in [12]. The purpose of a class `Grid` is visualization of different geographic grids, while class `WebMap` defines a layer that acts as an interface toward OGC Web Map Service compliant.

An extensibility of a suggested model consists of an ability to further specialize hierarchy of a `Layer` and/or `FeatureService` classes. The only condition that newly derived classes must respect is the predefined interface for operations defined on layers, i.e., feature services.

GADL and *GinisWeb* geoinformation are mutually associated. GADL, as an XML language, describes GIS applications using elements that correspond to classes defined in the model. On the other hand, automatic configuration of application components implemented by *GinisWeb* framework (based on the geoinformation model) requires description of an application domain that is encoded using GADL. For configuration of instances, all classes defined by the model possess virtual method `Create` that as an input parameter receives reference to a corresponding XML element from a GADL description.

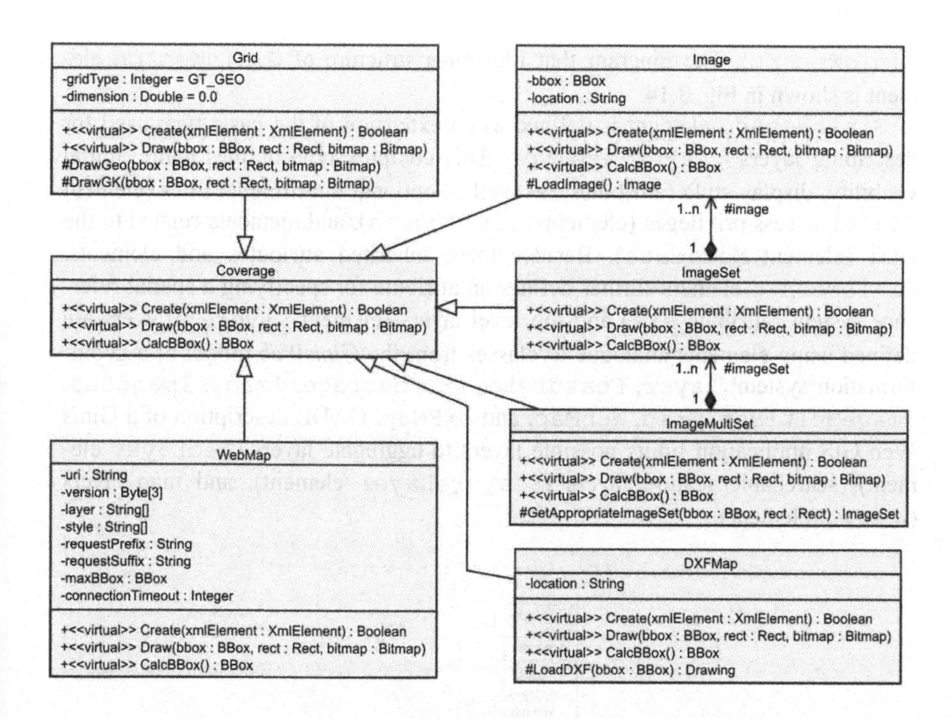

Fig. 3. 13. Detailed UML diagram of classes related to `Coverage` layers

4 Structure and basic elements of XML language GADL

Ginis Application Definition Language (GADL) is an XML language that is used for encoding an explicit description of a (Web)GIS application domain that is being built using *GinisWeb* framework. For the specification of GADL we used *XML Schema Definition Language* [4], so the definition of the language is an XML schema document[20].

XML language GADL is introduced to enable application describing, so new applications can be quickly developed. Unlike GML that is used to encode geodata, GADL is primarily used to encode data about geodata. GADL description of some GIS application, in essence, defines hierarchy and characteristics of layers, as well as corresponding feature services that are used as connectors to geodata sources. Web application that implements *Ginis Web GIS* represents, regarding to GADL description, a special case where sources of geodata are limited to WMS and WFS servers.

GADL schema specifies two separate root elements for describing *Web-enabled GIS node* (element `GinisApp`), and *Web GIS application* (element

[20] Complete GADL specification can be found at: http://gislab.elfak.ni.ac.yu/alexm/ginisweb.

GinisWebApp). The diagram that illustrates structure of GinisWebApp element is shown in Fig. 3.14.

GinisWebApp element is defined as an extension of the basic type used for describing layers - LayerBaseType. This complex type defines name, initial visibility, display style (attributes), as well as optional multilingual titles (element Title), access privileges (element Privilegies), and metadata related to the layer (element Metadata). Besides these inherited attributes and elements, GinisWebApp element further defines an attribute for specifying a spatial reference system (attribute srs) and top level layers (element Layers). Layers are defined using elements analogue to classes from the *GinisWeb* model of a geoinformation system: Layer, FeatureLayer, Coverage, Image, ImageSet, ImageMultiSet, Grid, WebMap, and DXFMap. GADL description of a Ginis Web GIS application limits possible layers to aggregate layers (WebLayer element), searchable feature layers (FeatureLayer element), and map layers (WebMap element).

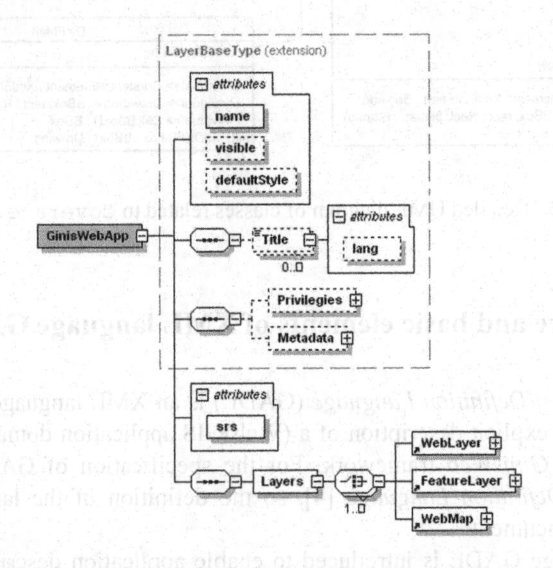

Fig. 3. 14. XML schema diagram showing the structure of a GinisWebApp XML element

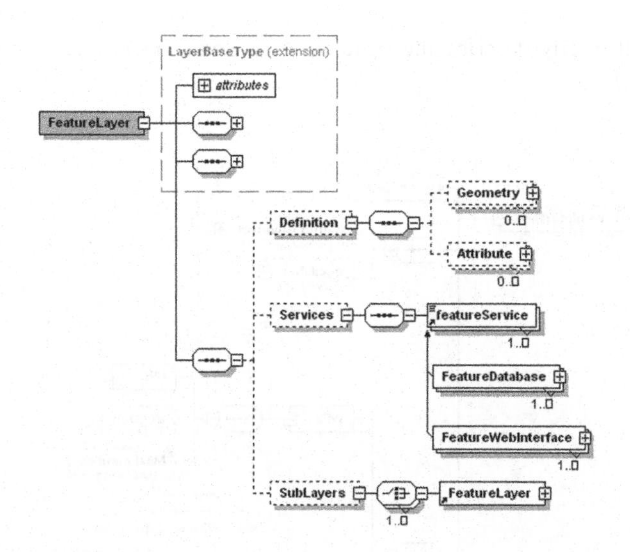

Fig. 3. 15. XML Schema diagram showing the structure of a `FeatureLayer` XML element

The detailed description of the structure of elements used for defining different types of layers is beyond the scope of this chapter. For the purpose of this GADL overview, the structure of `FeatureLayer` element will be described (Fig. 3.15). `FeatureLayer` is selected for its complexity and the fact that it can be contained by both Web-enabled and Web GIS descriptions.

Element `FeatureLayer` describes the layer containing features of specific type. Besides basic characteristics inherited from the previously mentioned `LayerBaseType`, this element can also contain `Definition`, `Services` and `SubLayers` elements. `Definition` element is used for describing non-spatial and geometric attributes that determine feature type, while `Services` element is used for describing services that provide and enable storing geographic features. The content of a `SubLayers` element in this case is reduced to a `FeatureLayer` element only. The relation between super and sub feature type is similar to the relation between super and sub classes in an object-oriented paradigm. It means that sub feature type inherits all attributes and services defined for a parent feature type.

Different types of feature services can be defined substituting abstract element `featureService` with the following elements: `FeatureDatabase` (describes connection to a geodatabase) and `FeatureWebInterfaces` (describes connection to WFS and/or WMS servers). `FeatureDatabase` service is typically used with Web-enabled GIS descriptions, while alternate `FeatureWebInterface` is used with Web GIS application descriptions. To complete this overview we will briefly describe the structure of a `FeatureWebInterface` element (Fig. 3.16).

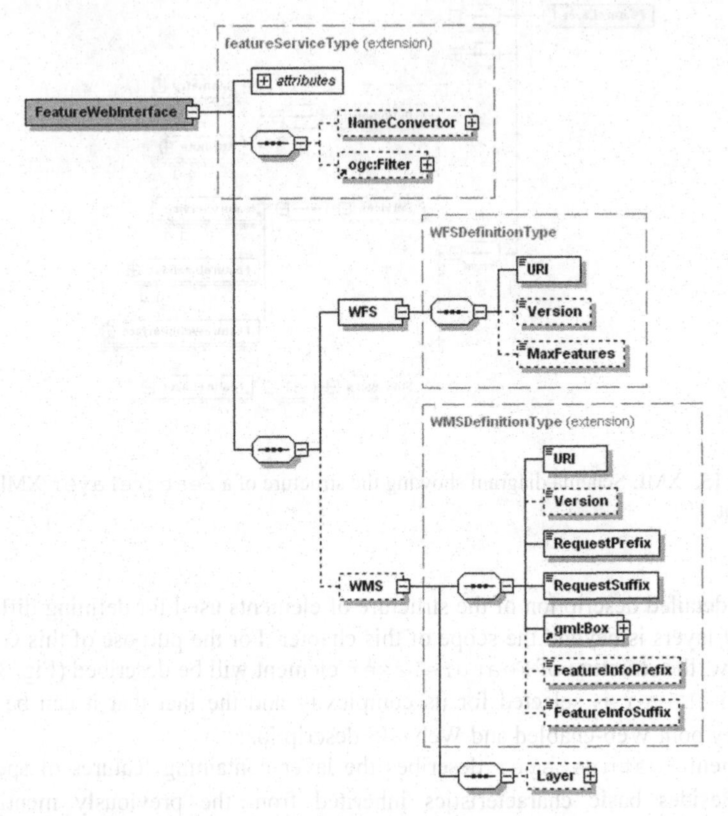

Fig. 3. 16. XML Schema diagram showing the structure of a `FeatureWebInterfaces` XML element

As defined in GADL, the basic complex type that copes with description of feature services is `featureServiceType` element. Through its attributes `insert`, `edit` and `delete` it allows us to define the use level of a service. Optionally, description of a service can specify conversion of names between an application and a service (`NameConvertor` element), and a filter for features that should be retrieved (imported element `ogc:Filter`). Elements specific to `FeatureWebInterface` are `WFS` and `WMS`. These elements are used for specifying parameters needed to enable successful connection to corresponding OGC services.

Finally, at the end of this overview we give two GADL examples (Fig. 3.17) concerning descriptions of a feature layer *Countries*. The type of features contained by the layer is defined with a single polygonal geometry named *Area*, and three non-spatial attributes: *Name* (string), *Population* (integer) and *Capital* (string). The first example (Fig. 3.17a) represents a segment extracted from some Web GIS application GADL description, while the second (Fig. 3.17b) can be a part of corresponding Web-enabled GIS description. The main difference between these two descriptions is the definition of a service. In the first case, the service is defined using `FeatureWebInterface` element that contains `WFS` and `WMS` elements specifying parameters needed for connecting with appropriate OGC compliant servers. As opposed to this, in the second example we have a geodatabase playing the role of a more powerful feature service that additionally enables insertion, editing and deleting. The difference between these two examples is also present on the level of details used for defining the geometry and attributes of the feature type. In the second example the `Geometry` and `Attribute` elements are enriched with read-only and required flags, default values and valid value ranges. This additional information is used by a desktop GIS application responsible for Web GIS content creation and maintenance.

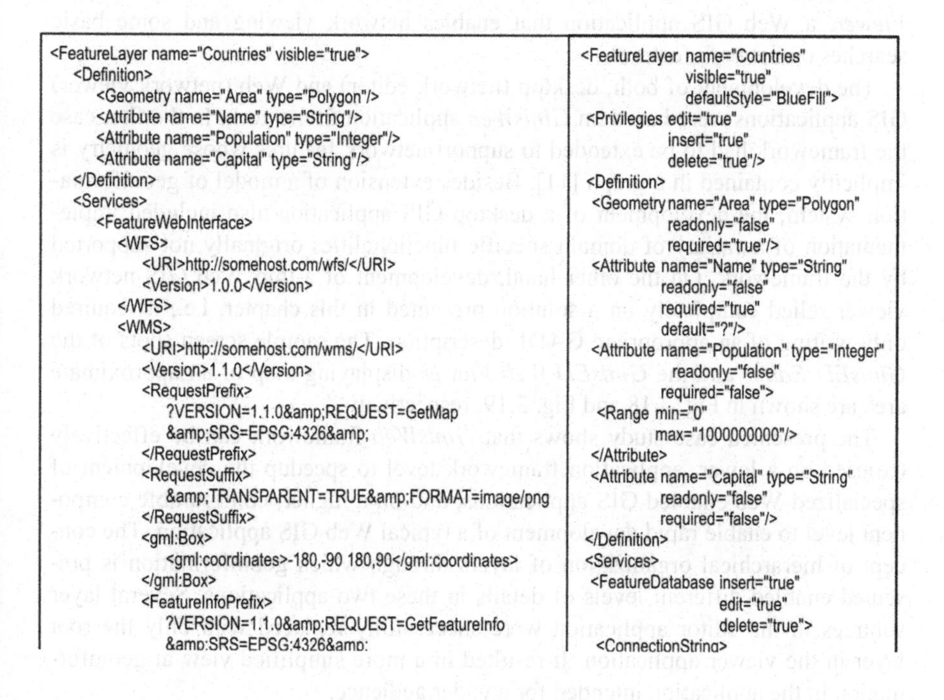

```
<FeatureLayer name="Countries" visible="true">
  <Definition>
    <Geometry name="Area" type="Polygon"/>
    <Attribute name="Name" type="String"/>
    <Attribute name="Population" type="Integer"/>
    <Attribute name="Capital" type="String"/>
  </Definition>
  <Services>
    <FeatureWebInterface>
      <WFS>
        <URI>http://somehost.com/wfs/</URI>
        <Version>1.0.0</Version>
      </WFS>
      <WMS>
        <URI>http://somehost.com/wms/</URI>
        <Version>1.1.0</Version>
        <RequestPrefix>
          ?VERSION=1.1.0&amp;REQUEST=GetMap
          &amp;SRS=EPSG:4326&amp;
        </RequestPrefix>
        <RequestSuffix>
          &amp;TRANSPARENT=TRUE&amp;FORMAT=image/png
        </RequestSuffix>
        <gml:Box>
          <gml:coordinates>-180,-90 180,90</gml:coordinates>
        </gml:Box>
        <FeatureInfoPrefix>
          ?VERSION=1.1.0&amp;REQUEST=GetFeatureInfo
          &amp;SRS=EPSG:4326&amp;
```

```
<FeatureLayer name="Countries"
              visible="true"
              defaultStyle="BlueFill">
<Privileges edit="true"
            insert="true"
            delete="true"/>
<Definition>
 <Geometry name="Area" type="Polygon"
           readonly="false"
           required="true"/>
 <Attribute name="Name" type="String"
            readonly="false"
            required="true"
            default="?"/>
 <Attribute name="Population" type="Integer"
            readonly="false"
            required="false">
  <Range min="0"
         max="1000000000"/>
 </Attribute>
 <Attribute name="Capital" type="String"
            readonly="false"
            required="false"/>
</Definition>
<Services>
 <FeatureDatabase insert="true"
                  edit="true"
                  delete="true">
  <ConnectionString>
```

Fig. 3. 17. Sample GADL segments describing a feature layer *Countries* from a Web GIS application perspective (a); and corresponding Web-enabled GIS perspective (b).

5 Case study: A Web GIS for an electric power supply company

The *GinisWeb* application framework has been successfully used for the development of several demo Web GIS applications[21]. Nevertheless, the case study presented in this chapter concerns its most complex application – the support of development of a GIS for capturing, maintaining and analysing an electric power supply network. This system has been built for the Serbian power supply company "Jugoistok".

When the project was started in 2005, the primary concern was the development of a desktop GIS that would enable digitalizing a power supply network based on a set of legacy paper maps. This resulted in an application named *GinisED Editor* and corresponding geodatabase. Later, in parallel with a process of digitalizing the network, the application has been extended with various functions related to certain domain specific analyses. Eventually, the need to widen data availability emerged. Our solution was the introduction of *GinisED Web Viewer*, a Web GIS application that enables network viewing and some basic searches on network features.

The development of both, desktop (network editor) and Web (network viewer) GIS applications were based on *GinisWeb* application framework. In the first case the framework had to be extended to support network features whose geometry is implicitly contained in a graph [11]. Besides extension of a model of geoinformation system, the development of a desktop GIS application also included implementation of a variety of domain specific functionalities originally not supported by the framework. On the other hand, development of a thin Web GIS network viewer relied completely on a solution presented in this chapter, i.e., it required only writing of an appropriate GADL description. The sample screen shots of the *GinisED Editor* and the *GinisED Web Viewer* displaying map of an approximate area are shown in Fig. 3.18 and Fig. 3.19, respectively.

The presented case study shows that *GinisWeb* framework can be effectively (re)used on a lower, application framework level to speedup the development of specialized Web-enabled GIS applications, and on a higher, configurable component level to enable rapid development of a typical Web GIS application. The concept of hierarchical organization of layers through which geoinformation is presented enabled different levels of details in these two applications. Several layer subtrees in the editor application were successfully replaced with only the root layer in the viewer application. It resulted in a more simplified view at geoinformation in the application intended for a wider audience.

---

[21] Some demo Web GIS applications can be found at: http://gislab.elfak.ni.ac.yu/alexm/ginisweb.

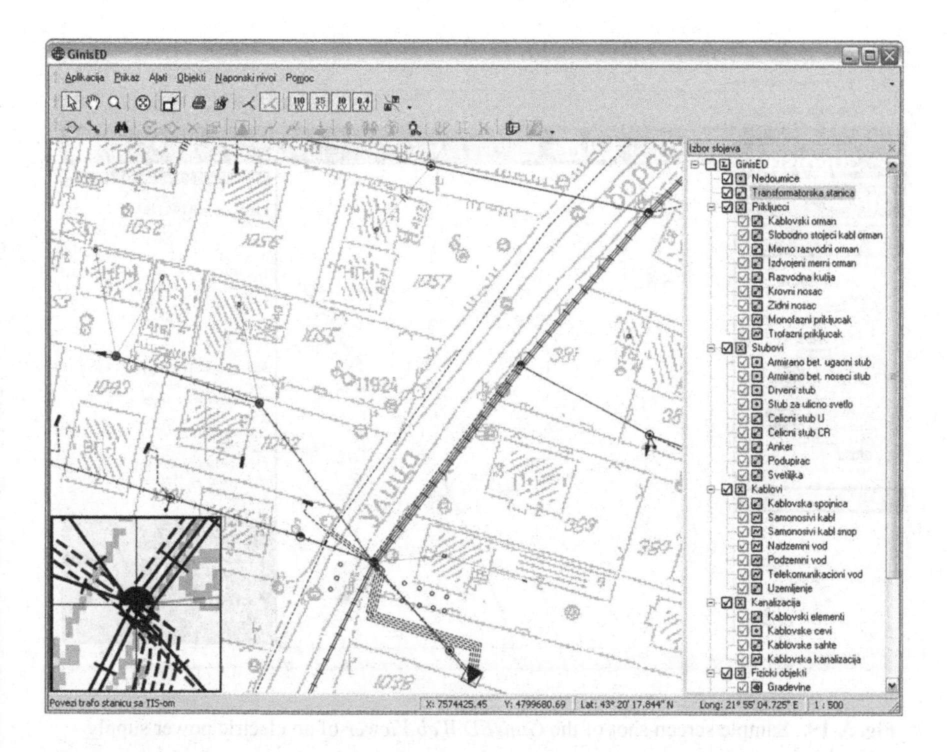

Fig. 3. 18. Sample screen shot from the *GinisED Editor* desktop GIS application for capturing, maintaining and analysing an electric power supply network.

6 Conclusion

This chapter introduces *GinisWeb* framework, a tool for a rapid development of an overall Web GIS. Rapid development is based on the use of configurable application components produced by the framework and on explicit XML descriptions of an application domain for their configuration.

Configurable application components of desktop GIS application, WMS, WFS, and Web application of a Web GIS are seen as different interfaces to a unique set of geodata and geoprocessing methods represented by *GinisWeb* model of geoinformation system. The model of geoinformation system represents the core for implementation of *GinisWeb* framework and the basis for specification of XML language GADL. In essence, GADL description of a domain defines layer hierarchy through which geodata is organized and corresponding services used for connecting with sources of geodata. GADL description of a Web GIS application is a special case where layer geodata sources have to be WMS and/or WFS compliant servers.

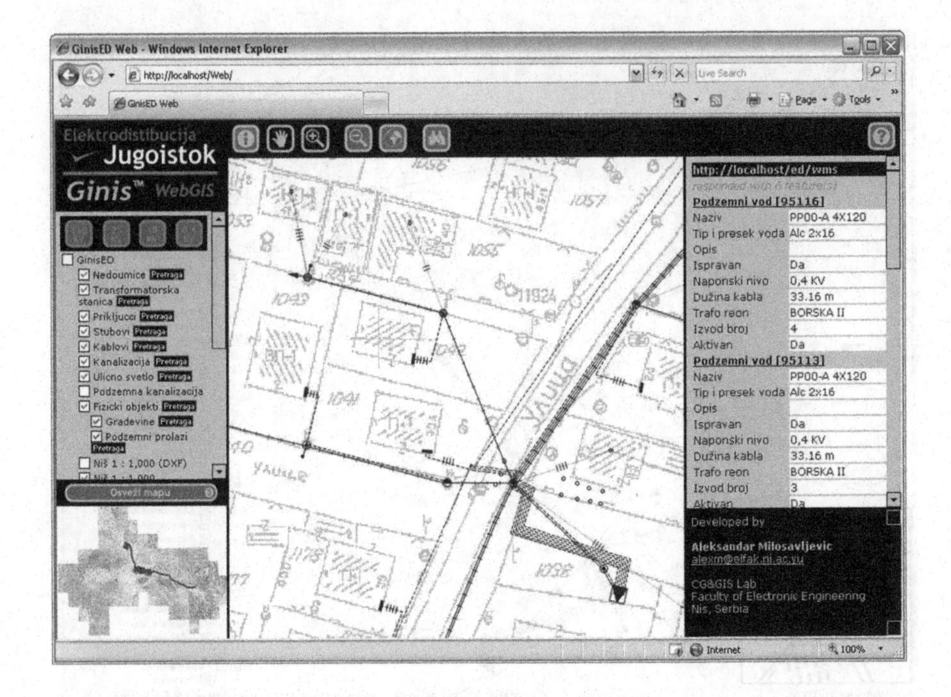

Fig. 3. 19. Sample screen shot of the *GinisED Web Viewer* of an electric power supply network.

The concept of configurable application components and XML descriptions for Web GIS development enables a high-level of reusability and extensibility, while openness is achieved through the adoption of widely accepted standards for communication between components. Lower levels of reusability are achieved through an application framework that allows us to introduce new types of geodata and widen geodata accessibility specializing `Layer` and `FeatureService` class hierarchy.

Acknowledgments This research was partially supported by the project "Geographic Information System for capturing, maintaining and analyzing of an electric power supply network", funded by Ministry of Science and Environment Protection, Republic of Serbia, and power supply company "Jugoistok", Serbia, Niš, Contract No. TR-6217A.

References

1. Buehler K (ed), OpenGIS Reference Model (Ver. 0.1.2), document 03-040, Open Geospatial Consortium, http://orm.opengeospatial.org, 2003.
2. Chang K, Introduction to Geographic Information Systems, Third Edition. McGraw-Hill, New York, NY, 2005.
3. Cox S, Cuthbert A, Lake R, Martell R (eds), OpenGIS Geography Markup Language Implementation Specification (Ver. 2.1.2), document 02-069, Open Geospatial Consortium, http://www.opengeospatial.org/standards/gml, 2002.
4. Fallside DC, Walmsley P (eds), XML Schema Part 0: Primer Second Edition, World Wide Web Consortium, 2004, http://www.w3.org/TR/2004/REC-xmlschema-0-20041028/
5. Fayad M, Johnson R, Schmidt D, Building Application Frameworks: Object-Oriented Foundation of Framework Design. John Wiley & Sons, New York, NY, 1999.
6. Kobryn C, Modelling Components and Frameworks with UML. Communications of the ACM, Vol. 43, No. 10, pp 31-38, 2000.
7. La Beaujardière J (ed), Web Map Service Implementation Specification (Ver. 1.1.1), document 01-068r3, Open Geospatial Consortium, 2002, http://www.opengeospatial.org/standards/wms
8. Mello UT, Xu L, Using XML to improve the productivity and robustness in application development in geosciences. Computers & Geosciences, vol 32, issue 10, pp 1646–1653, 2006.
9. Milosavljević A, Đorđević-Kajan S, Stoimenov L, An Architecture for Open and Scalable WebGIS. Proceedings of the 8th AGILE Conference on GIScience, Estoril, Portugal, May 26-28, pp 629-634, 2005.
10. Milosavljević A, Stoimenov L, Implementation Model of Ginis framework for Web-based GIS. ETRAN '05, Budva, Montenegro, Jun 5-10, pp 35-38, 2005.
11. Milosavljević A, Stoimenov L, Stojanović D, Dimitrijević A, Geoinformation model of a GIS for capturing, maintaining and analysing of an electric power supply network. YU INFO '06, Kopaonik, Serbia, March 6-10, 2006.
12 Rančić D, Đorđević-Kajan S, MapEdit: solution to continuous raster map creation. Computers & Geosciences, vol 29, issue 2, pp 115-122, 2003.
13. Shanzhen Y, Lizhu Z, Chunxiao X, Qilun L, Yong Z, Semantic and Interoperable WebGIS. Proceedings of the Second International Conference on Web Information Systems Engineering, Kyoto, Japan, vol 2, pp 42-47, 2001.
14. Shekhar S, Vatsavai RR, Sahay N, Lime S, WMS and GML based Interoperable Web Mapping System. Proceedings of the 9th ACM international symposium on Advances in geographic information systems, Atlanta, Georgia, USA, pp 106-111, 2001.
15. Soomro TR, Zheng K, Pan Y, Html and Multimedia Web GIS. Proceedings of the 3rd International Conference on Computational Intelligence and Multimedia Applications, September 23-26, pp 371-382, 1999.
16. Vretanos PA (ed), Filter Encoding Implementation Specification (Ver. 1.0.0), document 02-059, Open Geospatial Consortium, 2001, http://www.opengeospatial.org/standards/filter
17. Vretanos PA (ed), Web Feature Service Implementation Specification (Ver. 1.0.0), document 02-058, Open Geospatial Consortium, 2002, http://www.opengeospatial.org/standards/wfs
18. Wangmutitakul P; Minoura T, Maki A, WebGD: A Framework for Web-Based GIS/Database Applications. Journal of Object Technology, vol 3, no 4, Special issue: TOOLS USA 2003, pp 209-225, 2004.
19. Wong SH, Swartz SL, Sarkar D, A Middleware Architecture for Open and Interoperable GISs. IEEE MultiMedia, vol 9, issue 2, pp 62-76, 2002.

20. Worboys M, Duckham M, GIS: A Computing Perspective. Second Edition. CRC Press, Boca Raton, FL, 2004.

Chapter 4: Geospatial Web Services: Bridging the Gap between OGC and Web Services

Elias Ioup, Bruce Lin, John Sample, Kevin Shaw

Naval Research Laboratory

Mapping, Charting, and Geodesy

Stennis Space Center, MS

Andry Rabemanantsoa, Jean Reimbold

University of New Orleans

Department of Computer Science

New Orleans, LA

Abstract This chapter highlights the differences between Web Services defined by the World Wide Web Consortium (W3C) and the Open Geospatial Consortium (OGC). Several techniques for mapping or translating between W3C services and OGC services are reviewed. The challenges lie in integrating the services and clients which utilize these differing standards. Bridging the gap between these two standards requires not only translating the interfaces but also mapping functionality.

This chapter presents techniques for mapping and dividing OGC service capabilities into multiple SOAP services that better fit the W3C Service paradigm. A single OGC Service is split into multiple W3C Services, each representing a single geospatial data set. This mapping approach better reveals the functionality of the service and allows services to be composed as individual units. The individual services also provide improved geospatial metadata support for service discovery. Converting SOAP services into OGC services is a more difficult task. Included in this chapter is a discussion of the Naval Research Laboratory's Geographic Information Database (GIDB) Portal system and its approach to this problem.

1 Introduction

Web Services are software systems designed to facilitate machine-to-machine interaction over a network. While any system created to support this goal can be classified as a Web Service, the term usually refers to the standards for XML communication defined by the World Wide Web Consortium (W3C). One standard, the Simple Object Access Protocol (SOAP), defines client-server communication using existing Web transports. The Web Service Description Language (WSDL) is another W3C standard used to describe the interface of a SOAP-compliant service. W3C Services are an attractive option for service developers and providers because they are general enough to be used in any application domain and are broadly supported by the technical community.

Despite the benefits of W3C Services, the geospatial services community has developed independent specifications for geospatial data exchange. The specifications of these services are maintained by the Open Geospatial Consortium (OGC). Unlike a W3C Web Service, each OGC standard is designed to handle a specific type of data. Additionally, OGC services are widely supported by GIS applications which allow off-the-shelf use by ordinary GIS users. The service query/response protocols are standardized and service functions are fixed across all service instances. These features make OGC Services simpler to implement and consume than W3C Services.

Web Services and OGC Services are independently developed standards and are not directly compatible. Establishing interoperability between W3C Web Services and OGC Web services would benefit both communities. Web Service-based applications will benefit when able to access the data provided by OGC Services. Similarly, there are numerous non-GIS Web Services that contain location-aware data which could be integrated into an OGC client application. Differences between the two standards make interoperability difficult, but bridging this gap makes geospatial knowledge ubiquitously available to a much larger community of applications.

1.1 Related Work

There have been several efforts to resolve the incompatibilities between W3C and OGC Services. In "Wrapping OGC HTTP-GET and -POST Services with SOAP - Discussion Paper," the authors present an XML Schema for representing key-value pairs in XML [1]. Key-value pairs are the primary query mechanism for OGC services and are used either through HTTP-GET or HTTP-POST. Once wrapped in XML Schema defined XML, they can be conveyed through SOAP calls. Responses from OGC services are often binary (for example: a JPEG image) and so they are also wrapped in XML via MTOM.

Additionally, in "OWS 1.2 SOAP Experiment Report OpenGIS Discussion Paper," the authors describe their approach to providing interoperability between OGC Services and W3C services [2]. Specifically, they develop WSDL documents and SOAP services that mirror the functionality of the OGC Services: WMS, WFS, and WCS. They propose a standardized set of WSDL's to be used by anyone following this approach. This allows for SOAP interoperability, but it requires service providers to provide duplicate endpoints to the same service.

Both of these approaches solve the basic problem of wrapping or translating the query mechanisms, but they do not address all the key differences between OGC and W3C Services. Namely, OGC services are based on a two step process: getting the server's capabilities and then getting, putting, or modifying the data. Thus, with these approaches, a W3C Service user has to use three steps: getting the WSDL, getting the server's capabilities and then getting/putting/modifying the data. Also, a UDDI based search for the OGC based WSDL would not reveal the important capabilities of the server. As part of this chapter, we propose a method for resolving these issues.

1.2 OGC Services

OGC Services are popular among the geospatial community because they are designed to describe geospatial data and services. The standards are easy to implement and use, especially on the client side. As a result, geospatial services are rarely implemented with non-OGC standards.

The OGC includes standards for a large number of geospatial services. The most commonly used are the Web Mapping Service (WMS), the Web Feature Service (WFS), and the Web Coverage Service (WCS) [3, 4, 5]. WMS serves map layer images, WFS serves vector data encoded in Geographic Markup Language (GML), and WCS serves gridded data in multiple file formats such as GeoTIFF. The OGC continuously develops standards for geospatial services which have not yet gained wide acceptance.

The OGC standards are similar to Web Services in their reliance on HTTP for transport and XML to express service descriptions, requests, and in certain instances, data. Each OGC Service is a separate standard. While each standard has similarities, they are not compatible. Each service defines different request parameters, has different request types, and returns different data types.

The service description, or Capabilities Document, provides detailed geospatial metadata about the service in a standardized way. For example, this document defines the representation of the spatial bounds as well as the spatial reference system of the service.

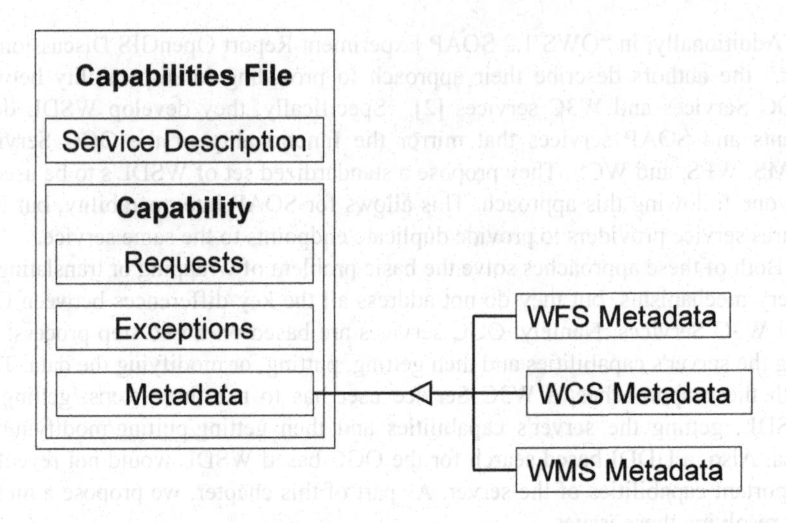

Fig. 4. 1. The OGC Service Capabilities Document includes metadata for the provided service: WCS, WFS, or WMS.

Possibly the most important feature of OGC Services is the uniformity of service functionality.

All services that implement a particular OGC standard will provide the same functionality. For example, every WMS server must provide a `GetCapabilities` function to retrieve its service description. A WMS server must also expose a `GetMap` function to retrieve a particular data layer. A standards-compliant client will be able to retrieve data from any OGC Service type by implementing support for these predefined functions. As a result, a completely general client can retrieve data from any OGC Service.

The main disadvantage of the OGC standards is their limitation to the geospatial community. Each standard provides a specific type of geospatial data or service. A general-purpose service not strictly providing geospatial data will be discouraged from using any OGC standard even though its data is associated with location context.

1.3 Web Services

The Web Service standards created by the W3C are widely used in many application areas, especially business-to-business network communication. Web Services use SOAP as the messaging method [6]. The SOAP standard expresses messages in XML and uses HTTP as platform-neutral transport. Every W3C Web Service provides a Web Service Description Language (WSDL) document that defines its functionality [7]. The WSDL lists each function definition of the SOAP service, the input parameters, and output types. The SOAP and WSDL standards form the basis for W3C Web Services.

The generality of Web Services allows them to wrap any kind of task, such as performing a database search. The service's functionality is exposed to any consumer independent of any programming language. As a consequence, two Web Services may describe identical input parameters and return types, but have completely different functionality. In this case, each WSDL document does not include any metadata that distinguishes between the two services. Other than service names and parameter types, all information on functionality must be provided externally to the service.

Many tools exist which simplify the use of Web Services. Web Service tools can automatically create application code to interface with a given WSDL or SOAP service, or create a WSDL from a particular program object. The Web Service registry standard, Universal Description, Discovery and Integration (UDDI) is widely supported and therefore useful in locating existing Web Services. The OGC equivalent is in limited use, requiring ad hoc solutions to locate available services.

Beyond the fundamental standards, Web Service extensions define new capabilities for usability and security. For example, WS-Security and WS-Reliability extend SOAP for services where sensitivity of data or high-availability is required [8, 9].

2 Interoperability

Interoperability between OGC standards and Web Service standards is important to the continued growth and availability of geospatial services.

While geospatial services are usually implemented with the OGC standards, most other services are not. Many existing Web Services are not geared toward geospatial data, though their data often includes a location component. Without a standard interface or the metadata necessary to access it, a developer must gather this knowledge from the Web Service's WSDL and adapt the client software to this interface. Geospatial applications which already have embedded OGC clients would benefit from access to this data.

On the other hand, there are many applications which have already integrated Web Service clients. These applications may not be solely focused on geospatial functionality, but they would benefit by integrating additional geospatial data or functionality provided by an OGC Service.

The additional tools, functionality, and extensions that exist for Web Services may be desired or even required for use with OGC Services. For example, there may be a requirement that all services used by an application implement the functionality of the WS-Security standard. Unless an OGC Service can be integrated into a Web Service, it will be prohibited from the system. In another scenario, an OGC Service may need to be included in a UDDI registry, which would be impossible without some Web Service interface.

Integrating Web Services-based geospatial data with standard OGC clients is greatly useful for standard GIS analysis applications. An OGC interface provides any OGC-compliant client access to the geospatial portion of the data. Without OGC interoperability a client would have to be specifically created for the Web Service. The generality of OGC clients precludes this requirement.

If a Web Service provides geospatial data there may be a drive to simply convert it into an OGC Service, allowing a much wider array of geospatial applications access to the data. However, in many cases a Web Service may not be exclusively dedicated to geospatial data. Instead, the geospatial data is one component of a larger data set. Location is such an important component of many application domains that many currently available Web Services contain geospatial data which is not being exploited by the users. These Web Services already provide important application functionality which cannot be removed simply to extend support for geospatial data. The better solution is to enable interoperability: integrating an OGC Service to provide proper geospatial support while leaving the original Web Service unaffected.

3 Implementation Issues to Consider

Supporting interoperability between W3C Web Services and OGC Services is non-trivial. In fact, providing an OGC Service interface to a Web Service and providing a Web Service interface to an OGC Service are two different and complicated problems. The underlying issue is the predefined functionality inherent to the OGC Service standards and the lack of any predefined functionality in the Web Service standards. Bridging the gap between these two standards requires not only translating the interfaces but also mapping functionality.

3.1 OGC to Web Services

Providing a W3C Web Services interface to OGC Services is an easier task in supporting interoperability. We focus on a solution that uses as a Web Service Wrapper around an OGC Service. This wrapper provides all the functionality of the OGC Service but with a standard W3C Web Service interface.

Data handling is an important issue to resolve between the Web Service wrapper and OGC Service. Web Services use XML as the primary method of communication. All non-string data types must be handled via special means. OGC Services use a variety of data types not limited to XML. The three most common

OGC Services all request data using URL-encoded parameters in the HTTP request or in XML. Both methods translate easily into the Web Service messaging model inside a simple wrapper. The difficulty arises in creating the response message. Each of the three OGC standards returns different data types. Though WFS uses XML, WMS and WCS use binary types which must be specially managed for inclusion in SOAP messages.

Another important consideration for the OGC to Web Service translation is the mapping of functionality. Web Services specify different requests as functions which are specified in the WSDL. OGC Services have one request to perform operations (getting/modifying/sending data and other requests for metadata (capabilities document/type definitions). The operations of an OGC Service are provided through one "function": `GetMap` for WMS, `GetFeature` for WFS, `GetCoverage` for WCS. The precise dataset provided by an OGC Service is hidden within the Capabilities Document. There is no method of determining what data a service provides without calling the GetCapabilities function. How do we provide access to the functionality of the OGC Service through the Web Service, while also creating a usable Web Service that follows best practices?

Metadata is one of the most important parts of the OGC Service standards. Each standard requires geospatial metadata be added to the Capabilities Document. This metadata is crucial to its effective use by a client application. However, no standard for geospatial metadata exists for Web Services. The Web Service WSDL provides metadata in addition to function definitions. Removing the metadata from the Web Service would remove much of its usefulness. As a result, it is important to create an effective method of including metadata inside the WSDL.

3.2 Data Handling

Service Translation

The simplest method of data handling between the OGC and W3C services is to only return string data types from the Web Service. Obviously, this presents a problem for OGC Services that return images or binary files. One solution is to require that the client retrieves binary data from the OGC Service directly. The Web Service interprets the clients request and returns a URL-encoded request for the OGC Service. We call this method "Service Translation" because there is no direct communication between the Web Service and the OGC Service.

Service Translation works by creating a translation Web Service which is designed to create request URLs for OGC Services from a SOAP request. The client creates the SOAP request for data and sends it to the Web Service. The Web Service parses the request and creates an equivalent OGC Service request and encodes it into a URL. Then the URL is returned to the client in the SOAP response. The client must then use the URL to retrieve the data directly from the OGC Service.

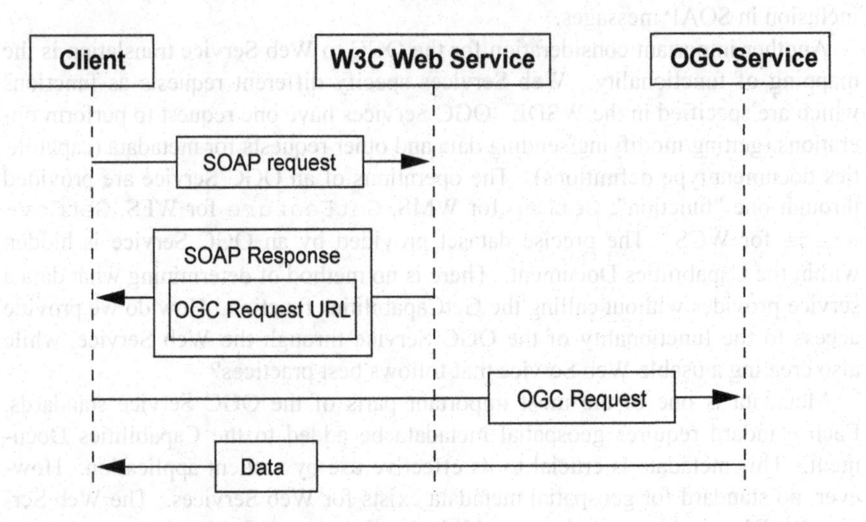

Fig. 4. 2. Service Translation reduces communication complexity, but clients must interact with both services in order to receive data.

The primary benefit of the Service Translation is that the Web Service does not have to manage messages containing binary data. Removing this functionality from the Web Service reduces the cost and complexity of communication. The Web Service does not have to act as a proxy for the OGC Service's data, removing the associated computational and network costs. However, the reduction in complexity and load on the Web Service are pushed to the client side. With Service Translation, the client must manage communication with both the Web Service and the OGC Service. While the client need not know the details of requesting data from an OGC Service, it still must make a second data request. Fundamentally, this is antithetical to the operation of Web Services. The goal of an OGC Service to Web Service mapping would be to remove as much complexity as possible from the client, a goal which loose coupling does not achieve.

In this type of system, a portion of the communication takes place outside of the Web Service framework. Thus, if Web Service specific extensions, such as WS-Security, are required for access, the translation prevents access to the service. The value of Web Services comes from operating within the Web Service framework and exploiting the functionality it provides. Bypassing this framework greatly diminishes this benefit.

This last disadvantage is the main reason we did not use this method in our interoperability system.

Service Wrapping

Service Wrapping is necessary to completely wrap an OGC Service with a Web Service. In this method, the data of the OGC Service is retrieved by the Web Service and then returned to the client. Service Wrapping will increase the load on the Web Service system and introduce complexities; however, it is in most cases the appropriate method of creating a Web Service interface to an OGC Service. Using a service wrapper will make all interaction with the client completely Web Service based. Thus, any specific Web Service requirements, such as use of WS-Security or WS-Reliability, will be possible. Clients for the Web Service can be made easily with existing tools and the system will mesh well in an existing Web Services infrastructure. Because wrapping meets the goal of completely hiding the OGC Service from the client and allows the use of all Web Service extensions, we chose to focus our work on this method.

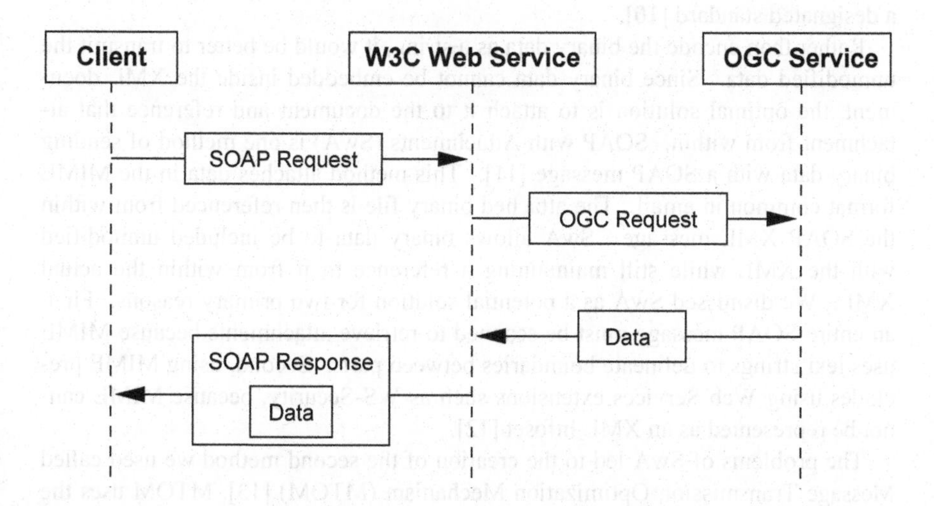

Fig. 4. 3. Service Wrapping incurs greater communication overhead, but simplifies the client tasking.

Service Wrapping requires that the Web Service be able to include binary data in its messages, specifically the response from a WMS or WCS server. The problem of including binary data with a SOAP response is not unique to geospatial service interoperability; thus there are some existing solutions for this problem.

However, none of these solutions provide the ease of access to data made possible by accessing the OGC Service directly.

Two methods that return data from a Web Service are available. The first is to encode the binary data in a string and return it within the XML. Base-64 encoding allows any binary data to be representing using only ASCII characters. The Web Service encodes the binary data from the OGC Service using base-64 encoding and returns it to the client. The client then must decode the base-64 data before using it. The main benefit to base-64 encoding of binary data is that the resultant string can be easily embedded inside the SOAP response. Any Web Service framework will be able to handle base-64 encoded binary data since it is functionally no different than standard string data. The problem with base-64 encoded data is that it is 33% larger than the original binary file. For large binary files often returned from OGC Services (such as large map images or GeoTIFFs) the increase in size will be significant. Encoding and decoding the base-64 messages will be computationally costly, especially if the service is high volume. The decoding task may have to be manually performed on the client side. While not difficult, it would be preferable to have the binary file available in its original form immediately, and delegate the binary data handling to the Web Services framework. Certain frameworks will automatically decode base-64 data, but this is not a designated standard [10].

Rather than encode the binary data as a string, it would be better to transmit the unmodified data. Since binary data cannot be embedded inside the XML document, the optimal solution is to attach it to the document and reference that attachment from within. SOAP with Attachments (SwA) is one method of sending binary data with a SOAP message [11]. This method attaches data in the MIME format common in email. The attached binary file is then referenced from within the SOAP XML message. SwA allows binary data to be included unmodified with the XML while still maintaining a reference to it from within the actual XML. We dismissed SwA as a potential solution for two primary reasons. First, an entire SOAP message must be scanned to retrieve attachments because MIME uses text strings to delineate boundaries between parts. Second, using MIME precludes using Web Services extensions such as WS-Security, because MIME cannot be represented as an XML Infoset [12].

The problems of SwA led to the creation of the second method we used called Message Transmission Optimization Mechanism (MTOM) [13]. MTOM uses the XML Binary Optimized Packaging (XOP) standard to include binary data in a file. All binary data is encoded using base-64 and included in the XML file. MTOM will package that XML document within an XOP package. All base-64 encoded data is removed from the XML and optimized, i.e. converted back to its original binary form. The binary data is still attached using MIME but within an XML Infoset that allows Web Services extensions such as WS-Security, which must compute signatures on the XML string data, to function properly. Using MIME allows MTOM to be backwards compatible with SOAP with attachments. MTOM retains compatibility with the Web Services model because of the temporary state where the data is base-64 encoded. At that point all data is in string representation and usable by any extension or tool which requires compatibility with the Web Services model. However, the transmission size is not inflated because the data is transmitted in the original binary format. The base-64 encoding of MTOM is also not a mandatory process; a client can access the original binary data from the message rather than having to base-64 decode it from a proper SOAP message. MTOM is a compromise which allows string-only representations of binary data without ever transferring the expanded form of the data. The main disadvantage of MTOM is that is must be supported by the Web Services framework to be fully successful. Because MTOM is a relatively recent standard there are Web Services frameworks which will not fully support it.

Our Web Service interface to OGC Services uses both base-64 encoding and MTOM as binary messaging methods. While base-64 encoding is not optimal, it will be supported by any Web Services system. The encoding and decoding procedure is well known and easily implemented. MTOM is too new a standard to enforce its usage. However, the reduction in transmission size is useful for our system which is geared toward heavy usage. As a result, we implement both methods of binary messaging in different functions, allowing the client to choose the method appropriate for their application.

3.3 Functional Mapping

Web Services and OGC Services have a fundamentally different design. Web Services are designed to have a flexible set of functions which are described in the WSDL for the service. OGC Services have a static set of functions but a flexible set of data. The data is described in the Capabilities Document for the service. There are two different ways to map functionality between OGC Services and the Web Service wrapper, each useful in different contexts.

In the first method the WSDL of the Web Service lists the static function set of the OGC Services. For example, the WMS specification defines a `GetCapabilities` and `GetMap` function, leading the Web Service wrapper to have corresponding `GetCapabilities` and `GetMap` functions. The client would be required to call the Web Service version of the `GetCapabilities` function to fetch the metadata for the service and then call the `GetMap` function with the appropriate parameters. This is the method used Ionic Software by the OGC Interoperability Program [2]. This method is appealing because it precisely matches the process of receiving data from an OGC Service. Direct mapping is useful because it allows the same WSDL specifications to be used for all OGC Services of the same type. It also allows the use of a Web Service interface which is completely generic. Any provider of an OGC Service can plug-in a Web Service interface following this model without any code or WSDL modification. The direct mapping also allows OGC Services to be added to UDDI registries while still retaining the essential properties of the original OGC Service as done in OGC Interoperability Experiments [14]. However, directly mapping functionality does not match the Web Services model. For a Web Service, all functionality should be revealed within the WSDL. By directly mapping OGC Service functions into Web Service functions all the important information about what the service actually does is hidden. For this reason, we use an alternative method to creating interfaces that better follow the Web Service model.

Instead of a direct map between OGC Service and W3C Service functionality, we create a mapping between OGC Service data and W3C Service functionality. Rather than exposing a function, such as `GetMap`, the actual data layers are exposed, for example a layer such as "satellite imagery." The data layers can be exposed in two different ways. The first maps data layers of a single OGC Service into functions in a new W3C Service. The organization of the original OGC Service is left intact and the relationships between the data layers are still apparent.

The second method maps each data layer into a separate service. This method creates a large number of simple, atomic services. No relationships or organization implied by the original OGC Service exists within these separate services. With this approach, each new

W3C Service will have its own `GetMap` which returns the data from that particular layer. Each new service will implement a W3C Service port type containing a `GetMap` function and can be easily composed into W3C Service orchestrations with a Business Process Execution Language (BPEL) engine.

We chose the latter method of transforming OGC Layers into individual W3C Services. To accomplish this, we created a tool which automatically transforms an OGC Service into many W3C Services. First, a parser ingests the OGC Service capabilities document and determines the available layers. For each layer, the tool creates a WSDL that contains a `GetMap` function. This function will be used to retrieve the map data from the W3C Service. The inputs of the `GetMap` function are the same as the original OGC Service function except for the layer name which is hard-coded to the W3C Service. Existing automated tools (WSDL2Java) are used to create the W3C Service program code from the WSDL. The tool modifies this code to use an OGC Service wrapper library which performs the actual request translation and forwarding to the OGC Service. This system is completely automated, allowing the new W3C Services to be created without any intervention. The process is similar for the alternate method of creating one W3C Service with many functions except that only one WSDL is made with a function for each data layer. Metadata for each new service can be provided by an additional `GetCapabilities` function which forwards a portion of the OGC Service capabilities document; however, we use an alternate method discussed in a later section.

As an example consider an OGC Service with three map layers: "RoadMap," "SatelliteImagery," and "HybridMap." Our automated tool will find the three layers and create three separate WSDL files. The three WSDL documents will describe three new W3C Services: "RoadMapService," "SatelliteImageryService," and "HybridMapService". Each of these services will have a `GetMap` function with input parameters of geographic bounds, image size, etc. The WSDL documents will be used to create program code for the W3C Services. The code for the `GetMap` function is modified to use the OGC Service wrapper which will provide the actual responses to W3C Service requests. In the case of the "RoadMapService" all requests will be turned into WMS URLs with the `layername` parameter set to "RoadMap." All other parameters are contained in the W3C request and are encoded into the URL. The URL will retrieve an image which is then sent back through the W3C Service either base64-encoded or using MTOM. The newly created WSDL, auto-generated service code, and OGC Service wrapper library can then be deployed and used via any standard W3C Service mechanism.

3.4 Metadata

Metadata is important to the proper usage of any OGC Service. The Capabilities Document of an OGC Service contains the list of data sets available from the service as well as the geospatial parameters over which the data is defined. A particular map may only be available over a small portion of the globe. It may also be available in multiple spatial reference systems. Knowing these parameters is necessary to determine whether a certain piece of data is useful for a particular application. While the metadata specification is standardized in the OGC Service Capabilities Document, there is no standard method of providing it from a Web Service.

The Web Service can provide access to the Capabilities Document of the OGC Service. Any client may then use the data as if it had been obtained directly from the OGC Service. But the Capabilities document does not serve as the primary description document for Web Services. As such, all Web Service functionality based upon the use of WSDLs will be missing vital information about the OGC Service. Moving metadata into the WSDL will provide Web Service-based tools and services with full information about a wrapped OGC Service.

The problem is that a WSDL normally does not contain metadata beyond the functions provided by the service and the parameter/return types of those functions. WSDLs are extensible. The metadata for the OGC Service can be included in the WSDL, but not in a standard way. While there are a limitless number of ways to encode metadata in a WSDL there are a few goals that should be met.

The first is that the metadata should not interfere with the proper usage of the WSDL. Any tools which do not use the metadata should not be affected negatively by it. Secondly, metadata methods should support all OGC Service metadata and be consistent between different service specifications. And finally, the metadata encoding method should allow for simple validation of parameters using existing XML tools.

Our method for providing metadata encodes it inside the extensible portions of the WSDL. Extensible information is only allowed in certain portions of the

WSDL. We include it inside the `<service>` element. The limits on the input parameters are encoded using an XML Schema definition with the XML Schema `<restriction>` element. Each input parameter is given a schema type which then has restrictions specified. For example, we can specify that the latitude of the request must be between 30 and 40 degrees or that the width of the return image must be less than 1000 pixels. The use of XML Schema to define capabilities provides a simple method of checking input parameters to the service. The input parameters can be validated against the restriction schema using a standard XML validation tool.

We add the metadata specifications for the response to its schema. Metadata elements are then placed within the response element in the WSDL. For example, the function may only return JPEG images. To signify this we add a format element with the string "jpg" to the response element. A client can parse these metadata parameters to determine the capabilities of the Web Service. Our method meets the three goals for encoding metadata in a WSDL. Information is only placed in the extensible portions of the WSDL, thereby not interfering with usage of the WSDL with any Web Service tool. The XML schema can encode all input restrictions defined in OGC Services and also allow simple validation of input parameters.

This method allows us to capture the information from a `GetCapabilities` document in the service WSDL. By encoding metadata in the WSDL, we preserve the OGC two-step process of getting a service's capabilities and then executing the service. Consider the following example of a layer defined in a WMS capabilities document.

```
<Layer>
     <Name>RoadMap</Name>
     <Title>Road Map</Title>
     <SRS>EPSG:4326</SRS>
     <LatLonBoundingBox SRS="EPSG:4326"
          minx="-180.0" miny="-90.0"
           maxx="180.0" maxy="90.0"/>
     <BoundingBox SRS="EPSG:4326"
          minx="-180.0" miny="-90.0"
          maxx="180.0" maxy="90.0"/>
</Layer>
```

The following XML Schema snippet defines an element "RoadMapRequest" and encodes restrictions on the values of the input parameters. These elements represent the definition of the input parameters to a SOAP service and would be encoded in the WSDL or a referenced schema document.

```
<xsd:element name="RoadMapRequest">
     <xsd:complexType>
          <xsd:sequence>
               <xsd:element  name="bbox"
```

```
                        type="geotypes:BoundingBox"/>
                   <xsd:element name="size"
                        type="geotypes:Size"/>
             </xsd:sequence>
       </xsd:complexType>
  </xsd:element>

  <xsd:complexType name="BoundingBox">
     <xsd:sequence>
       <xsd:element name="latMin" type="xsd:double"/>
       <xsd:element name="latMax" type="xsd:double"/>
       <xsd:element name="lngMin" type="xsd:double"/>
       <xsd:element name="lngMax" type="xsd:double"/>
      </xsd:sequence>
  </xsd:complexType>

  <xsd:element name="specification">
    <xsd:complexType>
      <xsd:all>
       <xsd:element name="latMin"
           type="latRestriction"  minOccurs="0"/>
       <xsd:element name="latMax"
           type="latRestriction"  minOccurs="0"/>
       <xsd:element name="lngMin"
           type="lngRestriction"  minOccurs="0"/>
       <xsd:element name="lngMax"
           type="lngRestriction"  minOccurs="0"/>
      </xsd:all>
    </xsd:complexType>
  </xsd:element>

  <xsd:simpleType name="latRestriction">
        <xsd:restriction base="xsd:double">
       <xsd:minInclusive value="-90.0"/>
       <xsd:maxInclusive value="90.0"/>
          </xsd:restriction>
  </xsd:simpleType>

  <xsd:simpleType name="lngRestriction">
          <xsd:restriction base="xsd:double">
        <xsd:minInclusive value="-180.0"/>
        <xsd:maxInclusive value="180.0"/>
          </xsd:restriction>
  </xsd:simpleType>
```

An alternate solution would be to use WS-MetadataExchange. This would allow metadata about the service to be encoded as WS-Policy documents. While this adds a third step back to the process, it does follow a W3C standard methodology [15].

4 W3C to OGC

There are instances in which one would want to convert W3C Services to OGC Services. Most GIS applications (both Web-based and desktop-based) have built-in capabilities for handling OGC services, specifically WMS, WCS and WFS. Therefore GIS users can access those services with no need for code generation (automatic or manual). However, because of the generality of W3C Services no GIS applications have built-in capabilities for automatically ingesting those services. Thus it is desirable to consider how to provide an OGC Service interface to a geospatial Web Service.

W3C services are more general than OGC services, thus representing W3C services as OGC services can require a reduction or change of the functionality of the original W3C service. For example, consider the elevation data service provided by the United States Geological Survey (USGS). It is a typical W3C service with a WSDL and SOAP endpoints. The service is available at http://gisdata.usgs.net/XMLWebServices/TNM_Elevation_Service.php.

The service provides the "getElevation" service defined by this XML Schema snippet:

```
<s:element name="getElevation">
  <s:complexType>
    <s:sequence>
      <s:element minOccurs="0" maxOccurs="1"
         name="X_Value" type="s:string" />
      <s:element minOccurs="0" maxOccurs="1"
         name="Y_Value" type="s:string" />
      <s:element minOccurs="0" maxOccurs="1"
         name="Elevation_Units" type="s:string" />
      <s:element minOccurs="0" maxOccurs="1"
         name="Source_Layer" type="s:string" />
      <s:element minOccurs="0" maxOccurs="1"
         name="Elevation_Only" type="s:string" />
    </s:sequence>
  </s:complexType>
</s:element>
```

The function takes latitude and longitude coordinates as required parameters and returns the elevation at that location. Contrast this to typical OGC service calls which take a bounding box for the spatial component of the query. OGC users provide a bounding box which must be converted into one or more points for use with the USGS service. There are several ways to perform this conversion. The following figure illustrates three methods. First, we can simply use the center point of the box for our Web service query. Second, we can use the corners. Or finally, we can generate an evenly spaced grid of points in the box.

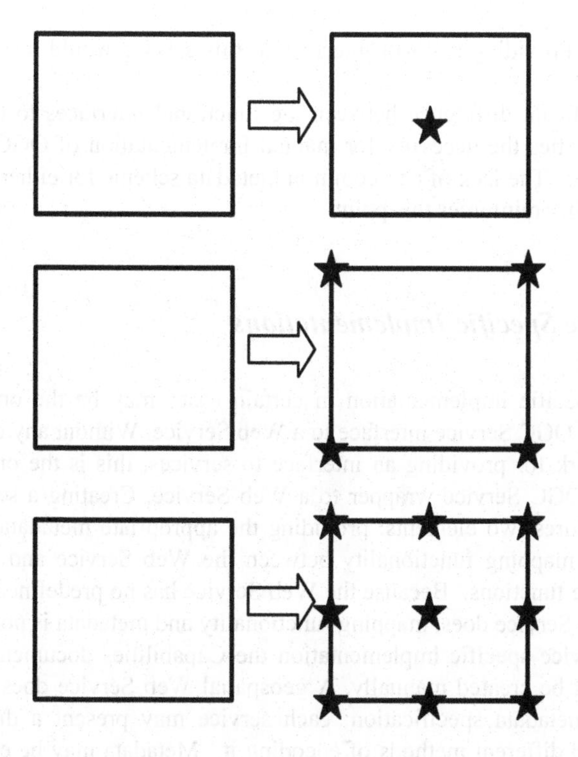

Fig. 4. 4. Example methods for converting bounding box to single point queries

The response format varies. For WMS we could return an image with elevation labels drawn at the points and for WFS we could return a GML document containing point features.

This example is provided to illustrate that the types of challenges in converting a general W3C service to an OGC service. Decisions are required about how to reduce or alter the W3C services functionality to fit the OGC framework. While it may be possible to generate a wrapper tool which can automatically convert any W3C service to an OGC services, its unlikely to suit the requirements of all or even most users. A certain amount of manual implementation will be necessary.

Two existing Web Services further exemplify this problem. ESRI's ArcWeb (http://www.arcwebservices.com) provides access to maps from a Web Service. The functionality for retrieving maps described in the WSDL is similar to that in the OGC Web Mapping Service; only a simple conversion of parameters would be required in an OGC WMS wrapper for the ESRI service. Conversely, Microsoft's TerraServer (http://terraserver.microsoft.com) provides access to maps from a Web Service with completely different functionality from the OGC Web Services. Their TerraService provides access to maps precut into tiles with a fixed numbering scheme. Maps cannot be retrieved using arbitrary geographic coordinates or image sizes. Providing a WMS interface to this service would take considerably more work.

The significant difference between the functional interfaces to these two services exemplifies the necessity for manual implementation of OGC interfaces to Web Services. The lack of any common metadata scheme for either of these Web Services further reinforces this point.

4.1 Service Specific Implementations

A service-specific implementation in certain cases may be the only method of providing an OGC Service interface to a Web Service. Without any other preexisting framework for providing an interface to services, this is the only method of creating an OGC Service wrapper to a Web Service. Creating a service specific wrapper requires two elements: providing the appropriate metadata for the OGC Service and mapping functionality between the Web Service and the mandated OGC Service functions. Because the Web Service has no predefined functionality but the OGC Service does, mapping functionality and metadata is not automatic.

For a service specific implementation the Capabilities document of the OGC Service must be created manually. A geospatial Web Service does not have any predefined metadata specification; each service may present a different set of metadata and different methods of encoding it. Metadata may be encoded inside an extended WSDL format, provided by a service function, provided externally to the service, or not at all. Manual construction of the Capabilities document requires combining these various sources of metadata. In certain cases the implementer may have to manually determine metadata not already provided by the Web Service to complete the OGC Service standard.

Each OGC Service standard must have all functions implemented as well. If the Web Service has functions with the same parameters and return types this only requires a simple wrapper that converts the OGC Service request to the SOAP request. A WMS will often fall into this scenario. The service requires only a few parameters such as geographic bounding rectangle and image size which are usually provided by any map image service. In contrast, WFS requires that data be provided using Geographic Markup Language (GML). If the Web Service does not provide access to vector data using GML a WFS wrapper must convert from the native Web Service format into GML. In many cases these service specific wrappers will be difficult and time consuming to create. Since no standard for geospatial Web Services exists, each OGC Service wrapper will be unique, making this process extremely costly to perform for a large number of Web Services.

4.2 Driver-Based Mapping

There is no method of automatically mapping a geospatial Web Service into an OGC Service. However, it is possible to improve upon the service-specific implementation in instances where many different Web Services mapped to OGC Services. Instead of implementing a completely new solution for each Web Service, the portion of the wrapper which outputs an OGC Service can be standardized. The OGC Service output obtains the geospatial data from an intermediate format internal to the system. To provide access to a new Web Service, a driver is created which maps the Web Service to the intermediate format. The drivers are much simpler to implement than an entire wrapper for each Web Service. This method reduces the time to provide interoperability to many Web Services.

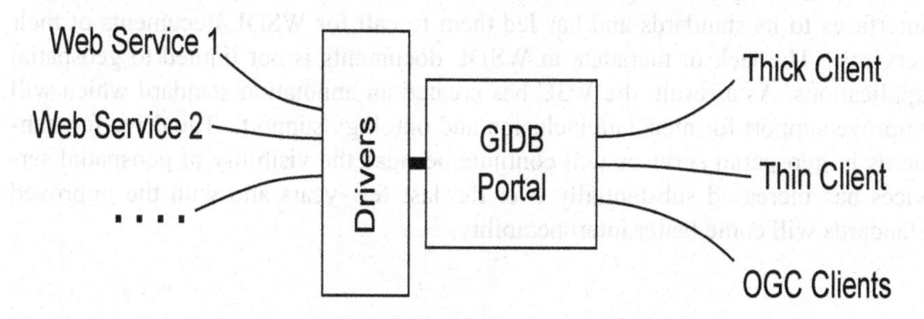

Fig. 4. 5. Driver-based interoperability

The Naval Research Laboratory's Geographic Information Database (GIDB) Portal system is an example of driver-based interoperability. We created the GIDB Portal to allow input from a variety of geospatial data sources, including Web Services. Each type of data source uses a driver designed to translate the data from remote source into the GIDB data model. The data sources in the GIDB data model are served out using a variety of interfaces. Raster data sources are served out as WMS layers and vector data sources are served as WFS layers. The GIDB data model contains all information necessary to automatically create a Capabilities Document for the OGC Service and to perform OGC Service queries for data.

5 Conclusion

The methods presented here provide the ability to unite OGC Services and Web Services. With the interfaces we have created, the extensive array of Web Service standards, tools, and extensions can be used with OGC Services. As the development of Web Services and the surrounding technologies continues to grow, the importance of providing a Web Service interface to OGC Service grows too. Our system provides a flexible system for providing this access while maintaining the important capabilities that are so desirable from OGC Services. By mapping data from the OGC Service to functionality in the Web Service we ensured that the capabilities of the OGC Service were fully described in the WSDL. On the other hand, our driver-based GIDB Portal is the simplest method of providing an OGC Service interface to a Web Service. While still a manual process, the driver-based system removes most implementation cost by reusing the system components that manage the data model and output data in OGC Service and other formats [16].

The task of creating interoperability between these two standards is difficult because of their respective strengths. The flexibility of Web Services standards and the strictness of OGC Service standards make interoperability difficult. However, this is likely to change over time. The OGC sees the need for Web Service interfaces to its standards and has led them to call for WSDL documents of their services. The lack of metadata in WSDL documents is not limited to geospatial applications. As a result, the W3C has created an annotation standard which will improve support for metadata inclusion and ontology support. The drive for standards in geospatial services will continue because the visibility of geospatial services has increased substantially over the last few years and with the improved standards will come better interoperability.

Acknowledgments The authors would like to thank the Naval Research Laboratory's Base Program, Program Element No. 0602435N for sponsoring this research.

References

1. Rüdiger Gartman and Bastian Schäffer. Wrapping OGC HTTP-GET/POST Services with SOAP. OpenGIS Discussion Paper, http://www.opengeospatial.org/standards/dp, January 2008.
2. Jerome Sonnet and Charles Savage. OWS 1.2 SOAP Experiment Report. OpenGIS Discussion Paper, http://portal.opengeospatial.org/files/?artifact id=1337, January 2003.
3. Jeff de la Beaujardiere. OGC Web Map Service Interface. OGC Implementation Specification, http://portal.opengeospatial.org/files/?artifact id=4756, January 2004.
4. Pangiotis A. Vretanos. Web Feature Service Implementation Specification. OGC Implementation Specification, http://portal.opengeospatial.org/files/?artifact id=8339, May 2005.
5. Arliss Whiteside and John D. Evans. Web Coverage Service (WCS) Implementation Specification. OGC ImplementationSpecification, https://portal.opengeospatial.org/files/?artifact id=18153, December 2006.
6. Nilo Mitra and Yves Lafon. SOAP Version 1.2 Part 0: Primer (Second Edition). W3C Recommendation, http://www.w3.org/TR/soap12-part0, April 2007.
7. Erik Christensen, Francisco Curbera, Greg Meredith, and Sanjiva Weerawarana. Web Services Description Language (WSDL) 1.1. W3C Note, http://www.w3.org/TR/wsdl, March 2001.

8. Anthony Nadalin, Chris Kaler, Phillip Hallam-Baker, and Ronald Monzillo. Web Services Security: SOAP Message Security 1.0. OASIS Standard, http://docs.oasis-open.org/wss/2004/01/oasis-200401-wss-soap-message-security-1.0.pdf, March 2004.
9. Doug Davis, Anish Karmarkar, Gilbert Pilz, Steve Winkler, and Umit Yalcinalp. Web Services Reliable Messaging (WS-ReliableMessaging) Version 1.1. OASIS Standard, http://docs.oasis-open.org/ws- rx/wsrm/200702/wsrm-1.1-spec-os-01.html, June 2007.
10. A. Ng, P. Greenfield, and S. Chen. A Study of the Impact of Compression and Binary Encoding on SOAP Performance. Proceedings of the Sixth Australasian Workshop on Software and System Architectures (AWSA2005), pages 4656.
11. John J. Barton, Satish Thatte, and Henrik Frystyk Nielsen. SOAP Messages with Attachments. W3C Recommendation, http://www.w3.org/TR/2000/NOTE-SOAP-attachments-20001211, December 2000.
12. Matt Powell. Web Services, Opaque Data, and the Attachments Problem. MSDN Web Services Technical Article, http://msdn2.microsoft.com/en-us/library/ms996462.aspx, June 2004.
13. Martin Gudgin, Noah Mendelsohn, Mark Nottingham, and Herve Ruellan. SOAP Message Transmission Optimization Mechanism. W3C Recommendation, http://www.w3.org/TR/soap12-mtom/, January 2005.
14. Josh Lieberman, Lou Reich, and Peter Vretanos. OWS1.2 UDDI Experiment. OpenGIS Interoperability Report, http://portal.opengeospatial.org/files/?artifact_id=1317, January 2003.
15. Keith Ballinger, Don Box, Francisco Curbera, Steve Graham, Canyang Keving Liu, Brad Lovering, Anthony Nadalin, Mark Nottingham, David Orchard, Claus von Riegen, Jeffrey Schlimmer, John Shewchuk, Greg Truty, and Sanjiva Weerawarana. Web Services Metadata Exchange (WS-MedatdataExchange), http://xml.coverpages.org/WS-MetadataExchange.pdf, February 2004.
16. J.T. Sample, R. Ladner, L. Shulman, E. Ioup, F. Petry, E. Warner, K. Shaw, and F.P. McCreedy. Enhancing the US Navys GIDB Portal with Web Services. IEEE Internet Computing, 10(5):5360, 2006.

Chapter 5: The Design, Implementation and Operation of the JPL OnEarth WMS Server

Lucian Plesea

Jet Propulsion Laboratory

4800 Oak Grove Drive

Pasadena, CA 91109, USA

Abstract The JPL OnEarth WMS server is a high performance WMS server offering public access to very large NASA earth imagery datasets, many of which have been created in conjunction with the server itself. This chapter provides description of the server implementation design and details of various components is provided, along with examples of how the components interact in actual use. The OnEarth server, in continuous service since mid 2000, has encouraged the development of a large number of client applications and has been an example of the benefits of interoperable GIS Web services.

As opposed to most WMS implementations, the OnEarth server only handles raster data, having no support for vector or point data. It is however a feature rich, scalable and fast implementation for WMS access to raster data, including support for OGC Styled Layer Descriptor (SLD). Most of the server performance is the result of the image processing pipeline implementation, which can use multiple CPUs while processing a single request, operates in memory and does not use any external temporary files. In addition to the main server binary, a tiled WMS and KML super-overlay is implemented as an apache module, offering a major boost in speed and availability to applications that can use pre-generated tiles. This component was proven capable of sustaining load exceeding two hundred WMS requests per second without overloading, while reducing the response time from a second or more to a few milliseconds. The WMS server also supports the concept of virtual image dataset, in which slave servers running on remote computers can be used instead of a local image dataset.

Web Map Service (WMS) is an Open Geospatial Consortium (OGC) standard protocol used for requesting and producing maps on the Internet, one of the main components of a suite of GIS interoperability standards. The WMS interaction between the client and the server is based on HTTP, with the client submitting requests in the form of Uniform Resource Locators (URLs)[22] which contain standardized parameters specifying the details of the operation requested, using either the POST or the GET method. Three operations are defined in the current WMS standard: a request for service-level metadata *GetCapabilities*; the actual request for a map *GetMap*; and an inquiry about a point on a map *GetFeatureInfo*. While the implementation of the GetFeatureInfo operation is optional, the GetCapabilities and GetMap operations are mandatory for the server. If specific map characteristics such as projection and format are supported by both a client and a server, these two standard operations are sufficient for an independent mapping client application to access the map. While the WMS standard offers many other capabilities, the simplicity of achieving basic WMS functionality contributed to the rapid success and proliferation of WMS, making it one of the most common Internet services, and making interoperable Internet based GIS applications a reality.

The JPL OnEarth WMS server, online at http://onearth.jpl.nasa.gov/, is one of the best known examples of a large scale implementation of a Web Map Service Server, providing GIS applications with easy access to a large collection of imagery and elevation data. The OnEarth WMS server became operational in early 2000 under the name of MapUS, enabling public access to one of the largest satellite images composites available at the time, a multi-spectral Landsat mosaic of the continental United States.

A precursor of the WMS protocol was used initially, making this server a contributor to many of the early OGC standardization activities. From this initial form, the OnEarth server data holdings and capabilities have continually increased. It is currently offering open WMS access to some of the largest mapping datasets available on the Internet. Many of these datasets have been prepared in conjunction with the server itself, with the explicit purpose of providing resources for the WMS and GIS user community by providing a data and feature rich WMS server.

While heavily utilized by many applications, the server complex itself has never transitioned into a stable operational status. Its main purpose is still limited to providing a rapid development platform and acting as a technology demonstrator. Secondary goals for the OnEarth server are to increase the awareness and the availability of NASA imagery data, acting in concordance with the NASA Mission Statement, "To understand and protect our home planet … To inspire the next generation … as only NASA can"

---

[22] Uniform Resource Locator: IETF RFC 2396

1. OnEarth Design

The OnEarth server design and its various software and hardware components will be explored in further detail, helped by a few examples illustrative of the server operations.

Unlike most WMS server packages, the OnEarth software is specifically targeted at handling large raster datasets, having very limited capabilities for vector or point data. For raster data however, it is a very flexible and feature rich yet scalable and high performance implementation of the WMS protocol. In addition to WMS, the OnEarth server also supports most of the OGC Style Layer Descriptor[23] (SLD) elements applicable to raster data and has support for high performance tiled WMS and native KML support. The SLD support greatly expands the WMS map representation options available to users, while the tiled WMS and KML capabilities makes it possible to support many concomitant users of interactive mapping applications. Both the software and hardware configuration of the OnEarth server have been developed in-house at JPL. The server development is driven either by specific user requests or as needed by new datasets.

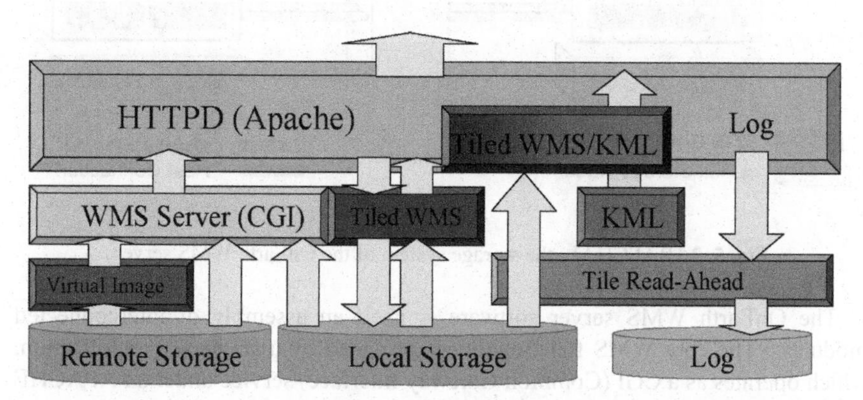

•Relative size is not significant •External Tiled WMS builds caches
•Only data flow is represented •Read-ahead prediction based on log

Fig. 5. 1. OnEarth System Diagram

The OnEarth hardware is represented by a heterogeneous collection of computers linked together by a gigabit Ethernet switch. The computers themselves are a dozen x86 Linux machines and a couple of medium size SGI IRIX servers. The Linux machines are mostly used for storage, ten of them forming a 40TB storage

---

[23] SLD Style Layer Descriptor, An OGC standard for specifying map presentations details. http://www.opengeospatial.org/standards/sld

system named RASCHAL (Raid Again Storage from Commodity Hardware And Linux), one being an input-output node and the last one providing storage for a large file download server.

The main OnEarth WMS server is hosted on an eight CPU SGI Origin 300 server; a smaller four CPU SGI Onyx machine is used for daily production of new images, while a third Origin machine with eight CPUs is used for large dataset development.

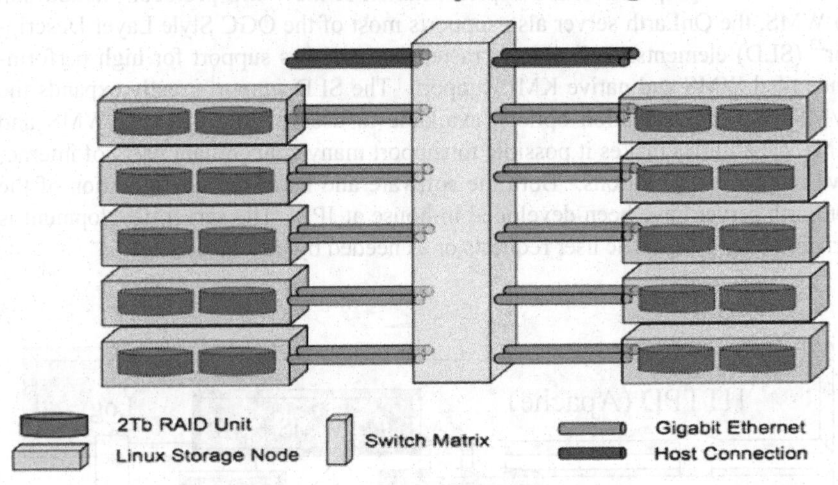

Fig. 5. 2. RASCHAL, the storage system of the OnEarth WMS server

The OnEarth WMS server software is itself an assembly of interconnected modules. The core WMS functionality is provided by a single C++ application, which operates as a CGI (Common Gateway Interface) service under an APACHE Web server. This application deals with the WMS request parsing and does most of the image processing required. The performance of this server module on the current hardware platform peaks at approximately ten WMS requests for the *global_mosaic* layer per second, while having an average latency of about one second. As detailed later on, a *global_mosaic* layer request triggers a complex processing chain on the server, requiring the cooperation of multiple computers. Requests for other layers are easier to handle so the server performance is slightly better. The efficiency and scalability of the WMS server application is the result of the processing model, which employs an in-memory, pipelined and multiprocessing application running on the SGI server. Since the WMS server spends most of the run time doing image processing, the server would not significantly benefit from using fast-CGI or any tighter integration with apache. It would actually make the implementation more complicated, introducing significant memory management and multiprocessing issues.

In addition to this main server, a fast access, tiled WMS and KML server is present, implemented as an apache module and serving pre-generated image tiles for specific client applications, without invoking the main WMS server. This module has proven capable of serving in excess of two hundred WMS tile requests per second while generating a minimal computational load, at the same time reducing the WMS response latency from about a second to a few milliseconds. It is limited only by the disk seek time and available network bandwidth. The KML support is itself tightly integrated with the tiled WMS implementation, automatically generating KML super overlay wrappers for all of the tiled WMS datasets.

The WMS server application also supports the concept of a virtual dataset, accessed via an Image Access Layer (IAL) subsystem. This system is capable of treating a remotely running IAL server the same as a local image file. In the current server configuration, this feature allows for preliminary image processing operations to be done by the storage computers which actually have the data stored on local disks, a configuration which simplifies the server application while at the same time eliminating some of the computational load from the main server.

Since the OnEarth WMS server has been in operation for a considerable amount of time with minimal interruptions in service, and is offering reasonable access to a rich and interesting set of data, a number of client applications have adopted the WMS protocol in general, and the OnEarth WMS server in particular, as an on-line source of geographical data. In the interoperability area, it is important to note that by choice the OnEarth WMS server development is not tightly coupled with any particular WMS client application, trying to be a general purpose mapping service provider.

2. OnEarth WMS Server

As described above, the core of the OnEarth WMS server is represented by a self contained CGI WMS application. This application has evolved significantly during the server existence, in a continuous effort to improve the performance and to add new features. The first implementation of the server was based on the MIT OrthoServer[24] code, which implements the Web Map Testbed protocol, a precursor of the WMS. This implementation used a combination of PERL and shell scripts for WMS request handling together with a few custom binary utilities for the map image processing.

---

[24] MITOrthoServer, http://ortho.mit.edu/orthoserver/

In conjunction with the creation of the WMS Global Mosaic[25], the OnEarth server code was completely rewritten and updated to support the WMS 1.1.1 and SLD 1.0 standards. The server code is written in C++, combining the system configuration and WMS protocol implementation with an image processing and formatting module. The current WMS component was written by Richard Schreyer from UCSB, during a summer student assignment at JPL. It makes use of libCGI[26] for the CGI functionality and Xerces-C[27] for the XML and DOM[28] capabilities. It implements all the required and many of the optional features of the WMS 1.1.1 and SLD 1.0 standards. This code was intended to function as a stateless CGI application, being instantiated for each and every WMS request, which makes the start-up and configuration time a significant concern. To reduce this overhead, the WMS server uses a single XML configuration file which contains all the information about the datasets being served and the externally visible server capabilities. The file structure is closely patterned after the WMS capabilities format, with additions that encapsulate local configuration items such as the local data files; capabilities for each layer and definition of predefined styles. The style description for each WMS layer also follows the SLD 1.0 format, again with a few extensions for the server configuration. This implementation choice reduces the server complexity, since the same code is used to configure both the system defined and the user styles.

The configuration file is read for every request and parsed into the corresponding DOM tree, making changes to the configuration instantly active. The WMS server supports both the HTTP POST and the HTTP GET mechanism, a feature of the underlying libCGI implementation. After the initialization and internal configuration, the WMS request itself is analyzed, and a *GetCapabilities* request simply retrieves the content of one of the configuration DOM tree nodes. If the request is a *GetMap* operation, the WMS parameters are checked against the configuration DOM tree and translated into a sequence of image processing requests, one such request for each requested layer, based on the internal server configuration. If a user SLD is requested, the respective style is parsed and merged into the server DOM tree, effectively becoming part of the configuration for the duration of this specific request.

The image layer processing requests are then passed to the image processing component of the WMS server binary via a list of arguments. This mechanism makes it easy to extend the image processing capabilities of the server without having to modify the WMS interface code. The extensions can be handled by

---

[25] WMS Global Mosaic, Landsat 7 multispectral mosaic, http://onearth.jpl.nasa.gov/WMS_GM.html

[26] libCGI: A CGI interface library written in C, http://libcgi.sourceforge.net/

[27] Xerces-C: A validating XML parser written in C++, http://xerces.apache.org/xerces-c/

[28] DOM: Document Object Model, an interface that allows an application to access and update the content, structure and style of XML documents, http://www.w3.org/TR/1998/REC-DOM-Level-1-19981001/

adding extra parameters in the server configuration file for a specific layer or style, parameters which are transferred to the image processing stage which is responsible for the interpretation. In the case of a WMS request that contains more than one layer, multiple instantiations of the map image are active at the same time, the results being combined using transparency blending.

If at any phase of the execution before the data processing an error is detected, the execution is interrupted and a relevant XML error message is sent back to the client instead of the expected image. Since the server was meant to encourage client development, the error detection is extensive, with the error messages being as accurate as possible in describing the cause of the errors and sometimes even suggesting the possible fixes.

The image processing component was independently developed as a standalone, geolocation aware image extraction and processing utility. It is still used as an independent application for data preparation. This component uses a multiple resolution, geolocated dataset representation as input, and is capable of subsetting, scaling, as well as having available a rich set of image processing operations. This application is based on the SGI Image Vision Library[29], which provides an extensible object oriented, multiprocessing framework for image processing.

This toolkit uses a demand-pull data flow model, in which a sequence of image operators is built by the application, with the processing itself is done only in response to an explicit output data request. This image output request propagates backwards in the processing chain until it can be satisfied. The processing itself is being done in arbitrary sized image regions; multiple processing tasks can be active at the same time in a multiprocessor environment. The OnEarth WMS server was designed to operate solely in memory, without the use of any temporary disk file. The multiprocessing synchronization and memory management required by this type of operating model is provided by the core framework. In addition to the processing capabilities available in the Image Vision Library, new image operators such as geolocation support and image storage, as well as data access facilities were custom implemented for the OnEarth WMS server. Multiple instantiations of the WMS image processing stage can co-exist, as is the case for a multiple layer WMS request. The final compositing of the output maps is done at the top WMS server level, followed by the final encoding of the resulting map image into the requested image format. As a continuation of the in-memory data-pull execution model, for complex and large WMS requests in a streamable output format such as JPEG or PNG, the sending of the first part of the formatted output image can start even before the image processing is complete, greatly reducing request latencies. The server can also generate GeoTIFF formatted map images, but since this format can not be streamed a temporary local file is used, making these type of requests somewhat slower.

---

[29] SGI ImageVision Library: C/C++ SGI toolkit for creating, processing and displaying images, http://www.sgi.com/products/software/imagevision

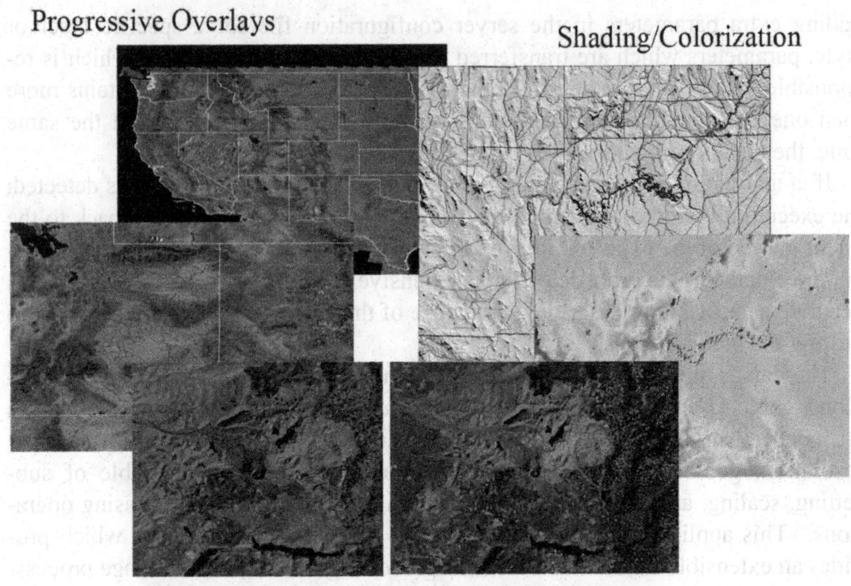

Fig. 5. 3. Examples of OnEarth image processing capabilities

The multiprocessing capabilities of the image processing framework and the efficient use of large memory space provided by SGI IRIX workstations are the most significant contributors to the OnEarth server performance. While the current server hardware is relatively modest, with only eight 600MHz CPUs and eight GB of memory, the server does scale to much larger configuration. This was demonstrated during the creation of the WMS Global Mosaic[30], when a similar code used multiple 32 CPU clusters with 16GB memory each, with very good efficiency. Indeed, in most cases the same data processing framework used by the WMS server is also used for the data preparation, making the integration of new data or features into the existing server relatively simple.

Even on the available modest hardware the performance is impressive, with latencies in the order of a second while handling about ten requests per second. Since a server overload can have serious consequences, a dynamic load limiting mechanism is built into the server. This mechanism attempts to keep the server utilization below a preset level by generating overload errors as replies to WMS requests as soon as the maximum server utilization level is reached. The overload condition is evaluated for each incoming request, the response to overload being very rapid.

---

[30] Remote Sensing and Supercomputing, A Match Made OnEarth: SGI User Group Meeting, 24th – 27th May 2004, keynote, Lucian Plesea, JPL

3. Pre-Tiled WMS and KML

One of the noticeable drawbacks of WMS is the amount of server side data processing required, the visible symptoms being large latencies and easily overloaded servers. While the OnEarth WMS server architecture is scalable and has a very low latency, the popularity of this service affected the server from the very beginning, overload being a common occurrence. For example, in just a few days after the test release of the WMS Global Mosaic became available on-line, attempts to download very large areas using the WMS protocol were already noticeable. For a short while, these attempts were managed by automatic blocking of the download clients or hosts.

However, the public release of the NASA WorldWind[31], which used the OnEarth server as the data source for both the WMS Global Mosaic and the SRTM elevation, drastically changed the server use pattern. Very rapidly the server computation load increased far above sustainable levels with request sources widely distributed across the Internet. The server was overwhelmed and access had to be restricted. At the peak, more than four million WMS requests were denied and logged each day, about forty times more than the server was capable of sustaining. Since resources for a hardware upgrade were not available, a software solution was the only possibility.

The requests from an interactive WMS map browser such as WorldWind follow a specific access pattern, programmed into the client application. The request are thus predictable, the required map images can be cached or created ahead of request time and then served from an existing database, eliminating the need for on-demand image processing. A WMS tiling system was implemented to handle this type of load. This OnEarth WMS tiling system implementation went through a few iterations, the current system being in the form of an apache server module. It was intended from the start to complement, not to replace the full WMS server. As such it is completely transparent to the client application. The tiled GetMap requests are indistinguishable from the regular WMS requests, making the two systems completely interchangeable. There is only one desired difference, the extremely rapid response of the tiled WMS and the absence of server overload errors.

Since becoming operational a couple of years ago, a number of such tiled WMS datasets have been added to the server for use in different client applications. Certain datasets such as the MODIS daily images are directly created in the tiled format, becoming simultaneously available to both the tiled and the normal WMS servers. This feature was enabled by making the tiled WMS storage format one of the underlying source image formats supported by the normal WMS server application, thus eliminating the need to duplicate the data and further integrating the tiled and the processing WMS capabilities.

[31] NASA WorldWind, A NASA open source virtual planet application.

The tiled WMS module itself is designed to operate as fast and efficient as possible. It uses features provided by the apache APR[32] library, and it was tested under many UNIX operating systems. This module is also using a single configuration file read during the apache server start-up, but as opposed to the WMS server, the tiled WMS module is a long running process, serving many tiled WMS requests. The incoming WMS requests are tested against the complete set of available tiled WMS patterns. To further reduce execution time, the test is done using pre-compiled regular expression patterns that contain the WMS parameters with the exception of the bounding box values. If a match is found, further checks are done to ensure that the resolution, size and location do match one of the available tile patterns, and the exact tile is selected. A request that does not pass these tests is ignored, with the apache server attempting the normal execution process, possibly invoking the full WMS server.

For a request that does match an existing tile, two sets of file open, seek and read operations are all that is required to retrieve the map image data and send it to the requestor. Since a single file is usually used to store the data for an entire layer, including multiple resolutions, the file open and file seek operations further benefit from the operating system caching. Furthermore, the OnEarth apache server runs in the *worker* multiprocessing model, under which multiple threads can responds to a very large number of requests, greatly reducing the server overhead. The WMS tiled module operating in this configuration has been able to handle in excess of twenty million requests per day, being essentially limited by the seek time of the media that stores the tile cache data. Very little CPU load is generated by this tiled WMS module, leaving most of the CPU cycles available for the full WMS server implementation. In an ad-hoc test due to hardware relocation, a laptop running the Linux version of this module and accessing the tiled data located on a remote NFS volume was able to provide tiled WMS service for a few days, with more than a million requests per day and without noticeable performance degradation when compared with the main server.

This tiled WMS server is already providing high performance service to many client applications, proving that no new data transfer mechanism other than WMS is needed. The only feature still lacking from the WMS standard is a method to expose such exiting tile patterns. This feature would simplify and automate the use of this service by client applications. To this purpose, a prototype tiling extension to the WMS protocol was implemented[33] on the OnEarth WMS server. This implementation adds a new operation type to the WMS protocol *GetTileService*, which is used to obtaining from the server information about the tiling capabilities. It is meant to be used in conjunction with the exiting WMS Capabilities, and as such it does not duplicate nor expand the layer or style metadata. The response to this request is simply an XML encapsulated list of the available WMS request

---

[32] Apache Portable Runtime: A set of libraries that provide a predictable and consistent interface to underlying platform-specific implementations.

[33] OnEarth Tiled WMS prototype: http://onearth.jpl.nasa.gov/tiled.html

patterns formatted as a URL, with the bounding box values being those for the top-left tile for each cached resolution. This mechanism provides a very simple way to expose the tiling, and can support any combination of WMS parameters, including layers, projections and SLD use. A client can simply ignore most of the parameters while still obtaining the information needed to request the tile images described by a certain pattern. A more sophisticated client can parse the pattern parameters in reference with the WMS server Capabilities metadata to further understand the nature of the map services and present this information to the user.

When different access patterns can be used to request the same exact data tiles, they are grouped together under a single *TilePattern* tag. The ability to deal with aliased requests was needed by OnEarth as a result of the multiple versions of WorldWind, where the developers modified the WMS data requests, adding, removing or reordering WMS parameters while still requesting the exact same map data. A hierarchical structure of tile patterns is used to further organize the tile patterns, the ones grouped under the innermost *TiledGroup* tag referring to different resolutions of the same map, while higher level *TiledGroup* tags are meant to be used by datasets related in some way, information meant to be presented to the user. A *Pad* parameter defined for a tile pattern allows for tile overlap, a feature useful in certain graphic applications. This tiled WMS model does not impose any tile size or alignment, and does not dictate a specific resolution sequence. While only recently implemented, a few issues with this mechanism are already known. For example, a mechanism for parameter substitution is needed to eliminate burdensome repetitions, for example the case of the Blue Marble Next Generation[34] dataset, which is offered in thirty six different styles, everything else being identical. Another example is the case of the tiled daily MODIS images, where the argument to the *time* WMS parameter is variable and determines the exact data to be served. Yet another known problem is the required precision of the coordinate values. While a real GIS client should have no problem providing the required precision, a simple image browser might not be so capable. One possible solution would be to add WMS tiling support for the map native coordinate system CR:1, where the coordinates are pixel locations in the base resolution. This alternative capability makes the coordinate location an integer value in all cases, and further exposes the raw map image details to the client.

To further improve the OnEarth server performance, a tile access predictor analyzes the Web access log in real time and if a tiled request is identified, it can issue system read commands for the surrounding tiles. If such a prediction is successful, a later tile request will already be in the system memory, making the response time even shorter by eliminating the disk latency.

---

[34] BMNG: http://www.nasa.gov/vision/earth/features/blue_marble.html

4. KML – WMS harmonization

The KML format is an XML encapsulation of geographical data which is used by Google Earth. Due to the popularity and availability of the Google Earth[35] application, this format is quickly becoming prevalent in the Internet mapping arena. While Google Earth natively supports an extremely limited set of the WMS standard, the Super Overlay mechanism introduced in the KML 2.1 version represents a very flexible extension mechanism for requesting map data, since the image data request can be any URL, including a WMS *GetMap* URL. Using the Super Overlay features, it is possible to provide KML wrappers for a set of tiled WMS requests. Furthermore, the URL requests for KML can themselves be WMS requests. This lead to the implementation of the KML to tiled WMS SuperOverlay translator which is currently integrated in the tiled WMS module itself, since all the data required for generation of the KML file is already available from the tiled WMS configuration file. When a KML WMS request is detected, a special handler routine converts the request to the matching WMS image request, and tests the existence of that specific WMS request within the tiled datasets. If a match is found, a KML file is assembled, containing the WMS request for image data, and if a higher resolution level of the specific WMS pattern exist, further requests for KML tiles, which will match the underlying WMS image tiles. Since all the data required for this operation already exists in the memory of the tiled WMS module, no disk access and no system calls are required for generating the KML wrappers, leading to a very efficient operation. Using this mechanism, all of the already available OnEarth tiled WMS datasets have become instantly available in GoogleEarth. The performance of the combined KML and Tiled WMS server is indeed very good, being faster than the static file implementation of KML SuperOverlays since no KML files need to be read from disk and the WMS image tiles are also very efficiently handled. Furthermore, the mechanism described above harmonizes the WMS protocol with KML, effectively transforming GoogleEarth into a fully featured WMS client, without requiring any changes in Google Earth itself.

5. Image Access Layer

The WMS server itself is built on top of a flexible image data access layer, which itself is derived from the SGI Image Vision Library. This data access layer provides uniform access to image data, presenting it to the WMS server itself as a three dimensional, multi-resolution and possibly multi-spectral georeferenced image data, hiding the exact storage details from the server. A number of custom

---

[35] Google Earth: http://earth.google.com/

image data readers have been implemented for this data access layer, the main ones being:

1. An indexed and tiled file format that can support extremely large images.
2. A client-server data reader that allows the use of a server residing on a different machine as a data source.
3. A composite image reader that allows the use of a collection of image tiles as a single large image.

It is easier to understand the role of these loaders by providing a few examples of their use in the OnEarth server environment.

5.1 Storage file format

The main data storage format is a tiled, multi-resolution and multispectral raster data storage format that is used extensively for the production of large datasets. In this format, data is organized in tiles of uniform size, each tile being read and written as a unit. The data tiles are stored concatenated in a single file, a second index file being used to organize and keep track of the location of each tile within the whole image space. Each tile data can be compressed using either lossless or lossy algorithms, with raw, JPEG (JFIF), abbreviated JPEG, zlib and bz2 being supported, together with a few unique optimizations for high byte count data types. A null index entry is used to symbolize non-existing data, providing explicit support for empty areas and providing efficient support for sparse datasets. Since the data and the index files can be specified separately, multiple index files can exist for a single data file, allowing different versions of a dataset to share the common areas.

The storage of the US elevation dataset can serve as an example of this feature. The data itself is stored as monochrome tiles of 16 bit integer values lossless compressed using zlib after a multiple byte optimization step. The value of each 16 bit signed integer is the elevation of a specific location expressed in meters. The same data file is used with a second index file, interpreted as a three value per pixel Hue-Saturation-Value (HSV) color image, where the Hue tiles are the elevation tiles for the specific location while all the Saturation and all the Value indices point to two constant value tiles, one for saturation and one for value. The end result in this case a color coded elevation image, an image of constant brightness and saturation, the hue of each point depending on the elevation of that specific location. In this way, the *us_elevation* and the *us_colordem* layers use the same data file with very different results, the only difference being the way the data is interpreted at the data access level and without any explicit support for this feature in the WMS server code itself.

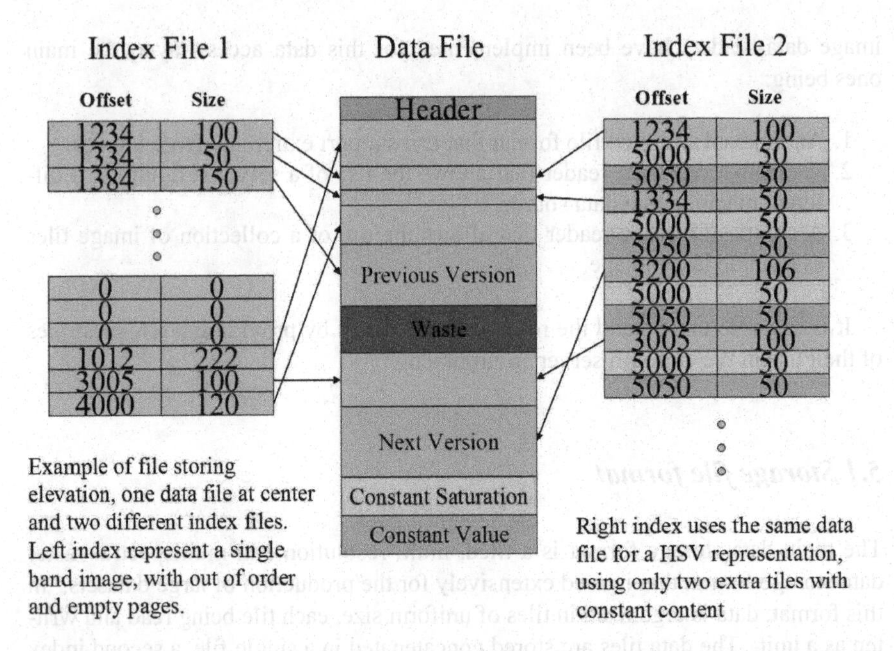

Fig. 5. 4. Detail of the file format used for storage, the US DEM case. The data file can be used with two different index files, being read either as elevation or as a color coded elevation.

5.2 Virtual Image Server

The dataset for which the current version of the server was explicitly designed is the WMS Global Mosaic, a full resolution, multispectral Landsat 7 mosaic of the GeoCover 2000 dataset, which covers most of the earth land mass.

The WMS Global Mosaic dataset contains nine spectral bands at different spatial resolutions. The highest resolution band is the panchromatic band, with a native resolution of half of an arc-second, or 2,592,000 by 1,296,000 pixels for the whole globe. The three visual bands and the three near infrared bands composites have a one second resolution, or 1,296,000 by 648,000 pixels. The last two bands are represented by the high and low gain thermal infrared, at two arc-second per pixel, or 648,000 by 324,000 pixels. When stored using the indexed and tiled format described above, with lossless compressed tiles and at the base resolution, the storage amount required is about 1.7 Terabyte. Over-sampling all the bands to the highest resolution and storing the result would increase the amount of raw data from 2.625 times the size of the panchromatic band to 9 times the size of the panchromatic band. This situation is handled by the OnEarth WMS server with the help of a virtual image server. This custom image server has direct access to three image files containing the WMS Global Mosaic bands stored in each of the three native resolutions. The server presents these data as a set of nine bands at uniform sampling of half arc-second. It does this by over-sampling the lower resolution bands as required. The operation of this server is completely transparent to the WMS server which in this case uses a network image reader to access this server instead of a local file reader. For example, when a WMS request for the *global_mosaic* layer using the *visual* style is received, the remote image reader is invoked. This loader opens a small local configuration file which contains the information needed to connect to the remote image server. To achieve better performance, four independent instantiation of the global mosaic image server exist, on two separate Linux storage units. Each unit has two server processes using a single local copy of all the bands of the WMS Global Mosaic dataset. This configuration was chosen based on the existing hardware, various other configurations being possible, with either more server hosts or server processes.

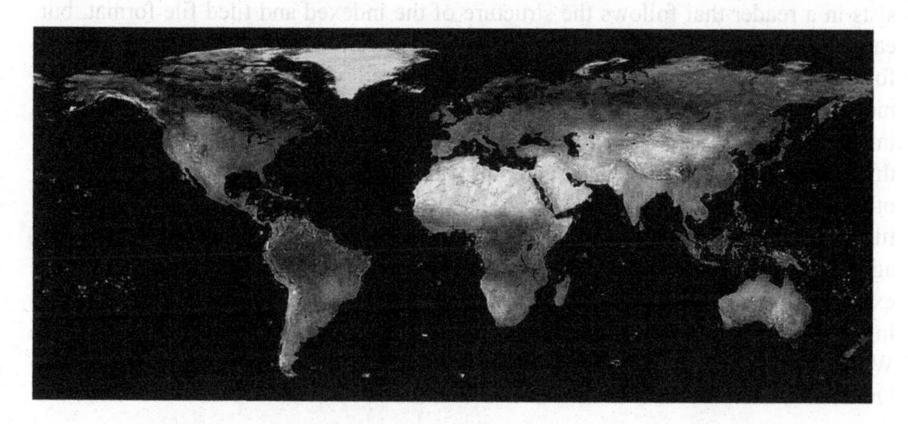

Fig. 5. 5. The WMS Global Mosaic, pan-sharpened pseudocolor

One of the available image servers is randomly selected and will be used for all the data used for producing the map image of this layer for this request. Once a connection is made, the data needed for this request is decomposed into a set of image tiles that mirror the existing storage tiles of the panchromatic band. For each such tile, four separate band requests are made to the image server, one for the panchromatic band and three for the each of the visual bands. If the requested resolution is half an arc-second, the one arc-second visual bands tiles that cover the requested area are read, decompressed and over-sampled to half an arc-second and the correct part of the resulting area is sent back as a response. This mechanism transfers some of the IO and computational load from the machine hosting the WMS server to the storage units that have direct access to the data, making it possible to process multiple WMS requests simultaneously without overloading the main server. The duplication of the global mosaic data not only represents a safety feature, but makes it possible to disconnect and operate on one copy while the WMS server continues to function using the other set.

The WMS Global Mosaic layers and styles are really virtual images, generated for each and every request and have never been stored as color images on the server. The common Landsat 7 pan-sharpening operation, which combines the higher resolution from the panchromatic band and the color information from the lower resolution bands is also available, being done by the WMS server image processing component.

5.3 Composite image reader.

This feature was mentioned when the tiled WMS was described, and simply consists in a reader that follows the structure of the indexed and tiled file format, but each individual tile is a fully formed image on itself, either JPEG or PNG. This format results in less efficient storage compared with the direct compression methods used by the index format, but presents two distinct advantages. These image collections can be treated as a single image, being possible to operate on this image collection with the same software tools and procedures used for any other file, including the full WMS server. It is also possible to read an individual tile and send it as a response to a WMS request without having to decode the image itself, feature which forms the basis of the fast-access tiled WMS module. For example, the daily MODIS composites available on OnEarth are directly created in this format using JPEG (JFIF) tiles. These files are usable immediately, as tiled WMS caches, and they are also available as a fully featured layer under the full OnEarth WMS.

6. Concluding remarks

Designing, building and improving the OnEarth WMS has been overall a great experience. The user feedback received is usually very positive, and in many cases it provides the only impulse to continue supporting or to add certain features of the OnEarth server. The reality is that due to very limited resources and the fact that the server is just a prototype and a development platform, many features and capabilities simply can not be made available.

Many of the features which are available on the OnEarth server are the result of a research effort to increase support for planetary GIS, the OnMars and OnMoon WMS servers being the concrete results of this effort.

One exciting project which is nearing completion at the time of this writing is the addition of a global and continuously updating image of the earth, built from the data collected daily by the NASA MODIS instruments. This image, at a resolution of 250m per pixel (about one sixth of a mile) would allow the continuous observation and exploration of the whole earth by anyone, the type of amazing applications made possible by the Internet.

It is very rewarding to see the varied and innovative ways in which the OnEarth services are used, and the increasing success of the interoperable GIS application model. The potential and realized impact of the OnEarth server have indeed been recognized by NASA, the OnEarth server complex and the various technologies described here have been the recipients of multiple NASA Space Act Awards.

Acknowledgments The research described in this (publication or paper) was carried out at the Jet Propulsion Laboratory, California Institute of Technology, under a contract with the National Aeronautics and Space Administration. The OnEarth server is the result of a long chain of projects and technologies, funded by various NASA projects and by JPL research and development. NASA Geospatial Interoperability Office sponsored the creation of the WMS Global Mosaic and the original implementation of the OnEarth WMS server. NASA ESTO-CT sponsored many of the image access, system and file storage technologies and the high performance computing aspects of the server. NASA AIST program sponsored development of planetary GIS, which resulted in the tiled WMS and KML implementation now used by OnEarth.

Chapter 6: Data Integration for Querying Geospatial Sources*

Isabel F. Cruz and Huiyong Xiao

Department of Computer Science

University of Illinois at Chicago

{ifc | hxiao}@cs.uic.edu

Abstract Geospatial data management is fundamental for many applications including land use planning and transportation and is critical in emergency management. However, geospatial data are distributed, complex, and heterogeneous due to being independently developed by various levels of government and the private sector. Until now, the formulation of expressive queries on geospatial data, which contrast with simple keyword-based queries, requires both user expertise and a great deal of manual intervention to determine the mappings between concepts in potentially dozens of data sources.

In this chapter, we describe an ontology-based approach to the problem of data integration, specifically focusing on the issue of query processing in a heterogeneous setting. Our contributions include a mechanism for metadata representation, an ontology alignment process, and a sound query rewriting algorithm for answering queries across distributed geospatial data sources. We demonstrate the practical impact of our approach in land use applications, which are exemplary of the extreme heterogeneity of data. We are leveraging current and emerging Semantic Web standards and tools for modeling, storing, and processing data. Our contribution to geospatial data integration is significant because new data sources can be added with relatively little effort, thus allowing for data manipulation and querying to extend seamlessly to the new data sources.

* This research was partially supported by the National Science Foundation under Awards ITR IIS-0326284 and IIS-0513553

1 Introduction

Years of autonomic and uncoordinated development of classification schemes by government organizations and the private sector pose enormous challenges in integrating geospatial data. In this chapter, our focus is on the integration of geospatial data that is created by the different counties and municipalities in the state of Wisconsin and stored locally by the counties and municipalities. Among the available geospatial data, we have concentrated on land use data. We have worked with the Wisconsin local government within the scope of WLIS (Wisconsin Land Information System) and the National Science Foundation under their Digital Government and Information Integration programs. Data heterogeneity in the land use data domain has been hindering the cooperation among the local governments to achieve comprehensive land use planning across the borders of the different jurisdictions [33].

We propose an ontology-based approach to enable integration and interoperability of the local data sources. In our work we deal with two kinds of ontologies: an axiomatized set of concepts and relationship types and a taxonomy of entities[17]. We call the first kind *schema-like ontologies*, because they are associated with the structure or *schema* of the local sources. The second kind model the entities (instance names) that describe land usage (for example, *agricultural*, *commercial*, or *residential*), where the only type of relationship is that of subconcept (or subclass) (for example, *single family residences* and *multiple family residences* are two subconcepts of *residential*), and are called *taxonomy-like ontologies* in our discussion.

The ontologies that we use to represent the structure of the local data sources, which we call *local ontologies*, belong to the first type and can be obtained from the source schemas through a schema transformation process. The second type of ontologies are *the land use ontologies*, which are part of the local ontologies; they represent land use taxonomies that classify land parcels in the local data sources according to their usage. In addition, our approach uses a *global* or *domain* ontology that models the domain associated with the task at hand and enables mediation across the local data sources. The global ontology contains a global land use ontology that describes the land use domain.

The key to our approach lies in establishing mappings between the concepts of the global ontology and the concepts of the local ontologies, a process called *alignment*. Using those mappings, a single query can then be expressed in terms of the concepts of the global ontology (or of a local ontology) and be automatically *rewritten* and posed against the other ontologies. We focus on database-style queries as opposed to simpler and less expressive keyword-based queries.

Query processing can be performed in two ways: *global-to-local* and *local-to-local*. In the former case, we rewrite a query posed on the global ontology into subqueries over the local sources (the global ontology acts as a uniform query interface of the integration system). In the latter case, we translate a query posed on a geospatial source to a query on any other geospatial source.

In this chapter, we consider the alignment process of the local land use ontology with the global land use ontology and propose an ontology alignment algorithm based on a set of deduction rules, which can be performed automatically when certain pre-conditions are established. We propose a *sound* query rewriting algorithm. The algorithm can compute a *contained* rewriting of a query in both global-to-local and local-to-local querying. Query containment ensures that all the answers retrieved by executing the rewriting are a subset of the answer to the original query, thus guaranteeing precise query answering across distributed data sources [23].

The rest of the chapter is organized as follows. The data heterogeneity issues in land use management are discussed in Section 2. The ontology creation process is described in Section 3. In Section 4, we focus on an automatic algorithm for ontology alignment. Query answering is presented in Section 5. In Section 6, we describe briefly the user interfaces that support ontology alignment and query processing. We summarize related work in Section 7. Finally, we draw conclusions and outline directions for future research in Section 8.

2 Data Heterogeneities

In this section, we describe in detail the kinds of heterogeneities that we encounter when integrating data from the local geospatial sources. In these sources, data is stored in XML format. Figure 6.1 shows two fragments of land parcel data, including their DTD (on the left-hand side) and an XML fragment (on the right-hand side), which respectively exist in the local systems of Eau Claire County and Madison County. As we can observe, even though the local XML sources present different structures and naming conventions, they share a common domain with closely related meanings (or semantics), thus being ideal candidates for an integration system.

The previous examples display *syntactic homogeneity* in that they both use XML but have different structures, therefore displaying *schematic heterogeneity*. They may also encode their instances or values in different ways, thus displaying *semantic heterogeneity*, in the sense that the same values may represent different meanings and that different values may have the same meaning [32]. Our discussion elaborates further on both kinds of heterogeneities. In the example shown in Figure 6.1, we see that the two source schemas overlap on most elements and both have the same nesting depth. However, the elements of the land use codes are represented differently in the two schemas: the schema S_1 uses four elements (broad, lu1, lu2, and lu3), whereas S_2 uses a single element (land_use). Furthermore, the values of such land use codes (in the XML instances) are encoded in different ways, namely characters for S_1 and numbers for S_2.

Land use codes in WLIS stand for land use types (or categories) and include, for example, *agriculture*, *commerce*, *industry*, *institutions* and *residences*. Besides using different names in different local source schemas, such land use codes have different classification schemes associated with them, thus resulting in semantic heterogeneities across the local source schemas. This is illustrated by Table 6.1, where there are four element names (Lucode, Tag, Lu1 and Land_use) from four different classification schemas. The descriptions in the table show that different values represent closely related land use types.

```
<?xml encoding="ISO-8859-1"?>
<!ELEMENT LandUse (LandParcel)>
<!ELEMENT LandParcel (AREA, BROAD, LU1,
  LU2, LU3, ..., JurisType, JurisName)>
<!ELEMENT AREA (#PCDATA)>
<!ELEMENT BROAD (#PCDATA)>
<!ELEMENT LU1 (#PCDATA)>
......
<!ELEMENT JurisType (#PCDATA)>
<!ELEMENT JurisName (#PCDATA)>
```

```
<LandUse>
<LandParcel>
<AREA>1704995.587470</AREA>
<BROAD>A</BROAD>
<LU1>AF</LU1>
......
<JurisType>County</JurisType>
<JurisName>EauClaire</JurisName>
</LandParcel>
......
</LandUse>
```

a) Local XML data source S_1 of Eau Claire County.

```
<?xml encoding="ISO-8859-1"?>
<!ELEMENT LandUse (LandParcel)>
<!ELEMENT LandParcel (AREA, LAND USE,
PARCEL ID, ..., JurisType, JurisName)>
<!ELEMENT AREA (#PCDATA)>
<!ELEMENT LAND USE (#PCDATA)>
<!ELEMENT PARCEL ID (#PCDATA)>
......
<!ELEMENT JurisType (#PCDATA)>
<!ELEMENT JurisName (#PCDATA)>
```

```
<LandUse>
<LandParcel>
<AREA>1007908.5</AREA>
<LAND USE>9100</LAND USE>
<PARCEL ID>246710</PARCEL ID>
......
<JurisType>County</JurisType>
<JurisName>Madison</JurisName>
</LandParcel>
......
</LandUse>
```

b) Local XML data source S_2 of the City of Madison.

Fig. 6. 1. Local XML land use data sources. In the data source S_1, BROAD and LU1 define the land use code, with BROAD as the first level and LU1 as a child level of BROAD. In the data source S_2 , LAND_USE specifies the land use code. In both sources, the elements JurisType and JurisName contain the jurisdiction type and name, respectively.

In our approach, a *local ontology* is generated for each local XML source that represents its schema. In addition, a *global* or *domain* ontology is defined to act as an integrated view and a uniform access interface to the distributed data sources. Every local ontology is mapped to this global ontology, by establishing the correspondences of their elements and attributes, which results in an alignment on the local names. In addition to this schema level reconciliation, it is also necessary to have a global land use taxonomy, to which the local land use taxonomies are mapped, so as to achieve a common understanding of the semantics of the land use codes in the local sources. All ontologies are represented using RDF and RDFS.

3 Ontology Creation

The first step of the integration of XML geospatial data sources is the transformation from the XML source schema and data to an RDFS ontology and to RDF data. This transformation encompasses the following steps:

Element-level transformation This transformation defines the basic classes and properties of the local RDFS ontology according to the transformation correspondences shown in Table 6.2, with the structural relationships between the elements not being considered for the time being. No new RDF metadata need be defined here because rdfs: Class and rdfs: Property are sufficient to express classes and properties. For instance, to transform the DTD of S_1

Table 6. 1. Semantic heterogeneity resulted from different encodings of land use data

Local Source	*Element Name*	*Land Use Type Value*	*Description*
DaneCounty RPC	Lucode	91	Cropland Pasture
Racine County (SEWRPC)	Tag	811	Cropland
		815	Pasture & Other Agriculture
Eau Claire County	Lu1	AA	General Agriculture
City of Madison	Land use	8110	Farms

in Figure 6.1, we define two classes: LandUse and LandParcel for the elements with the same name. The other elements become properties of LandParcel, because they are simple-type subelements.

Structure-level transformation This transformation encodes the nesting structure of the XML schema into the local RDFS ontology [11]. In particular, nesting may occur between two complex-type elements or between a complex-type element and its child (as a simple-typed element). Following the element-level transformation, the nesting structure in the former case corresponds to a *class-to-class* relationship between two RDFS classes, which are connected by the property rdfx: contained. In the latter case, the XML nesting structure corresponds to the *class-to-literal* relationship in the local ontology, with the class and the literal connected by the corresponding property. Table 6.3 lists the correspondences between the XML elements and the classes or properties in the local RDFS ontology.

Table 6. 2. Element-level transformation

XML Schema concepts	RDF Schema concepts
Attribute	Property
Simple-type element	Property
Complex-type element	Class

As an example, Figure 6.2 shows the local ontologies (represented as graphs where nodes are classes and edges are properties) transformed from the XML schemas in Figure 6.1. The land use taxonomies are transformed into a hierarchy of classes and incorporated as part of the local ontologies, rooted from LandUseTag and LandUseType, respectively.

4 Ontology Alignment

The ontology alignment process takes as input a local ontology and the global ontology and produces the class and property correspondences between them. We must consider two cases, which correspond to the schema and taxonomy components in the global ontology and local ontologies (see Figure 6.2): 1) *schema-level* mapping between the schema parts of two ontologies, where a concept (or a role) of one ontology is mapped to a concept (or a role) of another ontology, and 2) *instance-level* mapping, where two corresponding concepts use two different classification schemes for their instances, that is, land use codes with different underlying taxonomies in WLIS.

Ontology alignment is in general a challenging task, with its degree of difficulty depending on the types of ontologies being considered [30]. In our framework we have two kinds of ontologies and therefore two kinds of mappings, instance-level mappings and schema-level mappings. The former applies to two taxonomy-like ontologies consisting of only subClassOf relationships and the latter to two schema-like ontologies containing various properties and relationships (schema-level mappings). Next, we describe in detail the instance-level mappings and give an example with both schema-level and instance-level mappings.

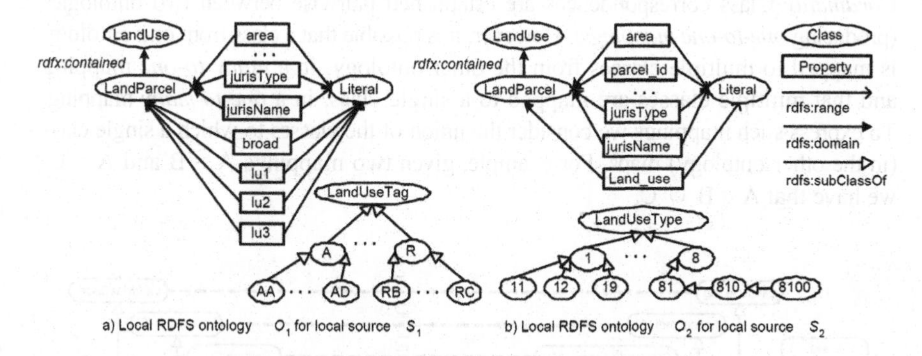

Fig. 6. 2. An example of local RDFS ontologies

Table 6. 3 Mappings between local XML schema D_1 and local RDFS ontology O_1

XPath expressions in D_1	RDF expressions in O_1
/LandUse	LandUse
/LandUse/LandParcel	LandParcel
/LandUse/LandParcel/AREA	LandParcel.area
/LandUse/LandParcel/BROAD	LandParcel.broad
/LandUse/LandParcel/LU1	LandParcel.lu1
...	...
/LandUse/LandParcel/JurisType	LandParcel.jurisType
/LandUse/LandParcel/JurisName	LandParcel.jurisName

Mapping types Figure 6.3 shows a fragment of two concrete land use taxonomies: the one on the left hand side is from the local ontology O_1 in Eau Claire County (as depicted in Figure 6.2) and the one on the right hand side is from the global ontology G.

The two taxonomies are respectively rooted from LandUseTag and from LandUseCode. A node in each taxonomy represents a class of land use, where the lable contains its description and the code (in parenthesis). The dashed lines represent the mappings that are established based on the semantics of the classes. We consider the following types in ontology mappings:

Semantic relationships Considering a set-theoretic semantics, the mapping between two classes A and B (seen as two sets of instances) can be classified into five categories: *superclass*, *subclass*, *equivalent*, *approximate* (or *overlapping*), and *disjoint*, respectively, $A \supseteq B$, $A \subseteq B$, $A = B$, $(A \cap B \neq \varnothing) \wedge (A - B \neq \varnothing) \wedge (B - A \neq \varnothing)$, and $A \cap B = \varnothing$.

Cardinality Class correspondences are established pairwise between two ontologies (producing *one-to-one mappings).* However, it is possible that a class from one ontology is mapped to multiple classes from the other ontology, in a *many-to-one* mapping and that multiple classes are mapped to a single class, in a *one-to-many* mapping. To express such mappings we consider the union of the classes to which a single class (in the other ontology) maps. For example, given two mappings A = B and A = C, we have that $A = B \cup C$.

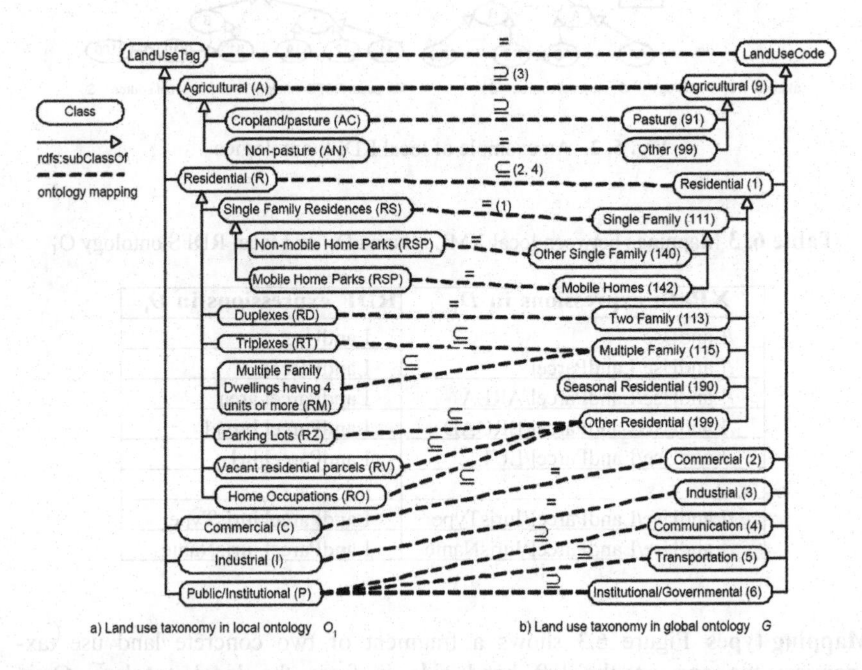

Fig. 6. 3. An example of mapping between two land use taxonomies. The labels over the edges represent mappings types, followed (in between parentheses) by the deduction rule(s) that can be applied, if any.

Coverage We distinguish two types of mappings: *fully covered* and *partially covered.* Let C and C' be two classes to be mapped, such that $C_1, ..., C_m$ are subclasses of C, and $C'_1, ..., C'_n$ are subclasses of C'. We say that C (resp. C') is *fully covered* if for each child $C_i \in \{C_1, ..., C_m\}$ (resp. for each child $C'_j \in \{C'_1, ..., C'_n\}$) there is a non-empty subset of $\{C'_1, ..., C'_n\}$ to which C_i is mapped (respectively there is a non-empty subset of $\{C_1, ..., C_m\}$ to which C'_j is mapped).

Deduction process In our approach, the ontology mapping process is performed using an inference process based on deduction rules. In the case that the deduction rules do not apply, then manual intervention by the user is needed.

This semi-automatic ontology mapping process follows two principles: (1) The deduction of the mapping between two nodes (from both taxonomies being mapped) is determined by the mappings between their children. In other words, the mapping between two ontologies are performed in a level-wise fashion, driven by the deduction rules that are defined based on the mapping semantics. (2) The user intervention is needed in two cases: when the mapping between two nodes has insufficient information to determine its type (for example, when some of the children of one node have not been mapped) or when there is conflicting information (for example, that a node is both a superset and a subset of the corresponding node).

We make the *complete-partition assumption:* for any class C in the taxonomy, its subclasses C_1, ..., C_n form a complete partition of the class, that is, $C = C_1 \cup ... \cup C_n$. For instance, in the global taxonomy depicted in Figure 6.3, the two children Pasture (91) and Other (99) of the Agricultural (9) class form a complete partition of Agricultural (9), since Other (99) includes all agricultural lands that are not used for pasture.

We consider the following deduction rules:

Definition 1 (Deduction rules). *Let C and C' be two fully covered classes, and C_1, ..., C_m and C'_1, ..., C'_n be the subclasses of C and C', respectively. Then, the mapping between C and C' can be obtained according to the following rules*:

1) $C = C'$, if for each $C_i \in C$, C_i is mapped to some k-element subset C'' of $\{C'_1, ..., C'_n\}(1 \le k \le n')$, such that $C_i = \bigcup_{l=1}^{k} C''_l$.

2) $C \subseteq C'$, if for each $C_i \in C$, C_i is mapped to some k-element subset C'' of $\{C'_1, ..., C'_n\}$ $(1 \le k \le n')$, such that $C_i = \bigcup_{l=1}^{k} C''_l$ or $C_i \subseteq \bigcup_{l=1}^{k} C''_l$.

3) $C \supseteq C'$, if for each $C_i \in C$, C_i is mapped to some k-element subset C'' of $\{C'_1, ..., C'_n\}$ $(1 \le k \le n')$, such that $C_i = \bigcup_{l=1}^{k} C''_l$ or $C_i \supseteq \bigcup_{l=1}^{k} C''_l$.

The deduction rules in Definition 1 can be proved to be *sound* and *complete* by an induction on the set-theoretic semantics of each rule, under the completepartition assumption and the assumption that the user-defined mappings are semantically correct.

The above rules assume a full mapping between C and C'. However, they still hold for the case of a partial mapping, provided that we define the following supplemental rule: *4) Suppose that a class C is partially covered by C' and that S is the subset of subclasses of C that are not mapped to any children of C'. Then, we create a temporary and empty subclass of C', and add a* superclass *mapping from each class in S to* .

In Figure 6.3, the symbols and numbers (in between parentheses) over the dashed lines (i.e., the class correspondences) indicate the mapping type and the adopted inference rule(s), respectively. For example, "⊆ (2,4)" over the mapping between the class Residential (R) and Residential (1) means that Residential (R) is a superclass of Residential (1), which is computed by rules 2 and 4. The application of rule 4 is due to the fact that SeasonalResidential (190) is unmapped, thus making Residential(1) partially covered.

Mapping representation The ontology mappings that result from matching the ontologies are stored in a file, called the *agreement* file. We use RDFS to express such mappings. Owing to the multiple inheritance feature of RDFS classes, the RDF property rdfs: subClassOf can be used in representing the three different kinds of mapping that may relate two classes *A* and *B*, namely $A \supseteq B$,, $A \supseteq B$, and $A = B$, in the taxonomy-like components of the two ontologies. For example, the first case is represented by the following RDFS segment:

```
<rdfs:Class rdf:about="A">
    <rdfs:subClassOf rdf:Class="B"/>
</rdfs:Class>
```

The second kind ($A \supseteq B$*)*, will be represented in the same way, by considering $B \subseteq A$. Finally, the third kind of mapping ($A = B$*)* will be represented simultaneously by the two different ways in which $A \supseteq B$ and $A \subseteq B$ are represented.

Regarding the mappings between the schema-like components of the two ontologies, in addition to relationships between classes, relationships between properties need to be expressed, namely *superproperty, subproperty,* or *equivalent (property)* mappings. Similarly to the previously described use of rdfs: subClassOf to represent class mappings, we can use the RDF property rdfs : subPropertyOf to represent these property mappings. Figure 6.4 shows a fragment of the RDFS representation of the mappings between the global ontology *G* and the local ontology O_1, which include the mappings between the schema-like and the taxonomy-like components of the ontologies.

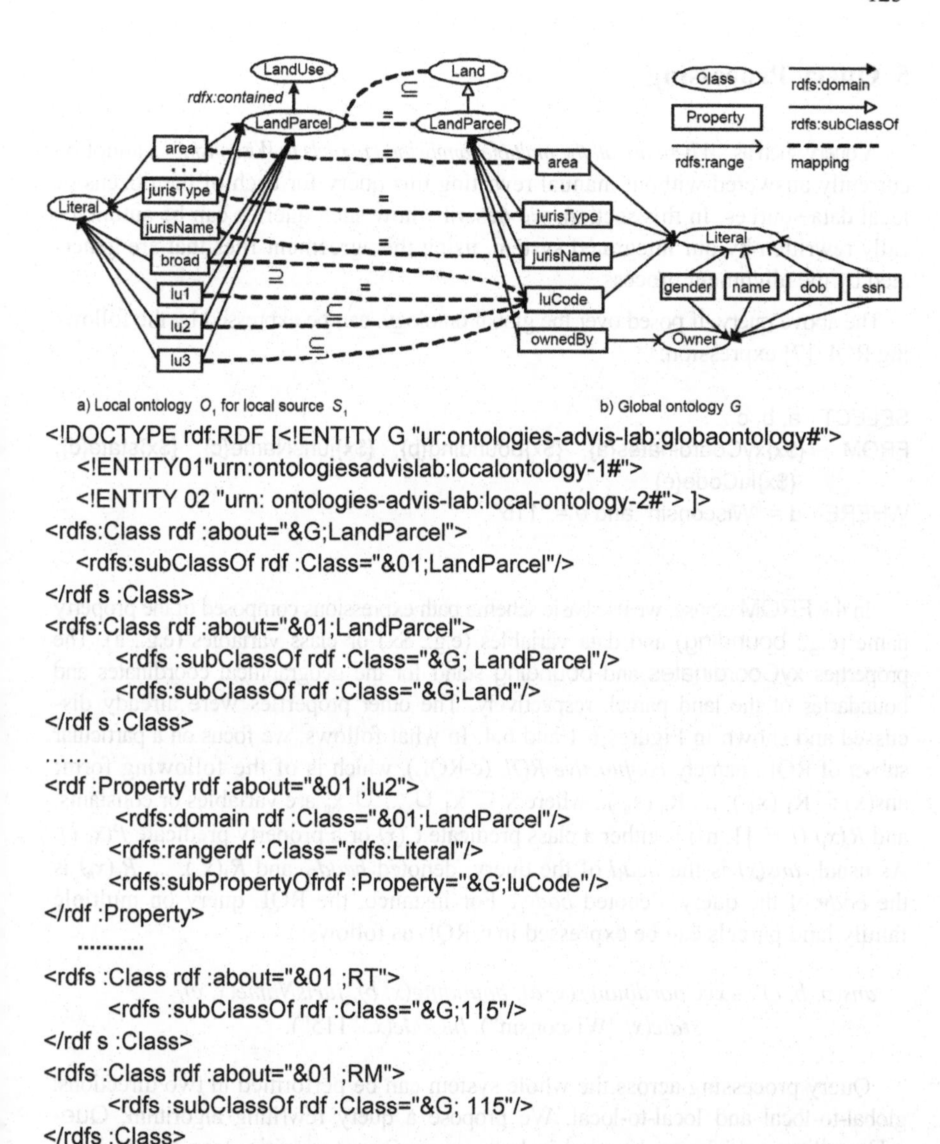


```
<!DOCTYPE rdf:RDF [<!ENTITY G "ur:ontologies-advis-lab:globaontology#">
  <!ENTITY01"urn:ontologiesadvislab:localontology-1#">
  <!ENTITY 02 "urn: ontologies-advis-lab:local-ontology-2#"> ]>
<rdfs:Class rdf :about="&G;LandParcel">
  <rdfs:subClassOf rdf :Class="&01;LandParcel"/>
</rdf s :Class>
<rdfs:Class rdf :about="&01;LandParcel">
     <rdfs :subClassOf rdf :Class="&G; LandParcel"/>
     <rdfs:subClassOf rdf :Class="&G;Land"/>
</rdf s :Class>
......
<rdf :Property rdf :about="&01 ;lu2">
     <rdfs:domain rdf :Class="&01;LandParcel"/>
     <rdfs:rangerdf :Class="rdfs:Literal"/>
     <rdfs:subPropertyOfrdf :Property="&G;luCode"/>
</rdf :Property>
  ...........
<rdfs :Class rdf :about="&01 ;RT">
     <rdfs :subClassOf rdf :Class="&G;115"/>
</rdf s :Class>
<rdfs :Class rdf :about="&01 ;RM">
     <rdfs :subClassOf rdf :Class="&G; 115"/>
</rdfs :Class>
......
```

Fig. 6. 4. A fragment of the agreement file as represented in RDFS. Local ontology O_1 uses a hierarchical land use code containing four properties: broad, lu1, lu2, and lu3, such that lu1 is a subclass of broad, lu2 is a subclass of lu1, and so on. In contrast, the local ontology O_2 only has one property, luCode, for land use coding. The mappings between the two ontologies are as follows: broad, lu1, lu2, and lu3 are respectively mapped to luCode, as a broader class (i.e., superclass), an equivalent class, a narrower class (i.e., subclass), and another narrower class.

5 Query Processing

A query such as "*Where are all the multiple family land parcels in Wisconsin?*" cannot be currently answered without manual rewriting this query for each of the dozens of local data sources. In this section, we describe how such queries can be automatically rewritten by our integrated system, using the agreement files that are generated by the alignment process.

The above query, if posed over the global ontology, can be expressed by the following RQL [7] expression:

```
SELECT  a, b, c
FROM    {$x}xyCoordinates{a}, {$x}bounding{b}, {$x}jurisName{c}, {$x}state{d},
        {$x}luCode{e}
WHERE   d = "Wisconsin" and e = "115"
```

In the FROM clause, we use basic schema path expressions composed of the property name (e.g., bounding) and data variables (e.g., $x) or class variables (e.g., a). The properties xyCoordinates and bounding stand for the geographical coordinates and boundaries of the land parcel, respectively. The other properties were already discussed and shown in Figures 6.1 and 6.4. In what follows, we focus on a particular subset of RQL, namely *conjunctive RQL* (c-RQL), which is of the following form: $ans(x) :- R_1(x_1), ..., R_n(x_n).$, where $x \subseteq x_1 \cup ... \cup x_n$ are variables or constants, and $R_i(x_i)$ ($i \in [1..n]$) is either a class predicate $C(x)$ or a property predicate $P(x, y)$. As usual, $ans(x)$ is the *head* of the query, denoted $head_Q$, and $R_1(x_1), ..., R_n(x_n)$ is the *body* of the query, denoted $body_Q$. For instance, the RQL query on multiple family land parcels can be expressed in c-RQL as follows:

$$ans(a, b, c) :- xyCoordinates(x, a), bounding(x, b), jurisName(x, c), state(x, \text{"Wisconsin"}), luCode(x, \text{"115"}).$$

Query processing across the whole system can be performed in two directions: global-to-local and local-to-local. We propose a query rewriting algorithm, QueryRewriting, which can be used in both cases. Query rewriting can be seen as a function $Q' = f(Q, M)$, where Q is the query to be rewritten, called *source query*, M is the set of ontology mappings, and Q' is the resulting query, called *target query*. The algorithm is shown in Figure 6.5.

In the global-to-local case, the source query Q is posed on the global ontology G, M is the set of mappings from G to every local ontology $O_1, ..., O_n$, and the target query Q' is the union of multiple subqueries over $O_1, ..., O_n$. In the local-to-local case, Q is a local query posed on a local ontology O_i ($i \in [1..n]$), M is the set of mappings from O_i to one or more local ontologies O_j ($j \in [1..n]$ and $j \neq i$), and Q'

is the union of multiple subqueries over all O_j. In the latter case, M is, in fact, a set of compositions of the mappings from O_i to G with those from G to O_j.

The QueryRewriting algorithm consists of four main steps: 1) *source query expansion* using the source ontology constraints, 2) *schema-level mapping* where the expanded source query is rewritten into an intermediate target query using schema-level mappings, 3) *intermediate target query expansion* using the target ontology constraints, and 4) *instance-level mapping* where the expanded intermediate target query is rewritten using instance-level mappings to obtain the final target query. In what follows, we cover the overall query processing by describing the three key components of the four main steps listed above: query expansion, schema-level mapping, and instance-level mapping. Finally, we discuss some of our assumptions and prove the correctness of the query rewriting algorithm.

Algorithm QueryRewriting (Q, M)
Input: a conjunctive query Q over ontology O; the mappings M between ontologies O and O'
Output: a union Q of conjunctive queries Q' over O'

---

1 head $_{Q'}$ = head$_Q$; body$_{Q'}$ = null;
2 Q* = QueryExpand(Q, $\sum$), where $\sum$ is the set of constraints over O;
3 Let ϕ be body$_{Q*}$;
4 Let M$_1$ be the part of schema-level mappings in M;
5 For each R(x) of ϕ
6   For each ψ $\in$M$_1$
7   Let R'(x') be the result of applying ψ on R(x);
8   body$_{Q'}$ = R'(x')$\wedge$ body$_{Q'}$;
9 Q' = QueryExpand(Q', $\sum$'), where $\sum$' is the set of constraints over O';
10 Let M$_2$ be the part of instance-level mappings in M;
11 Q = ConstantMapping(Q', M$_2$);
12 Return Q;

Fig. 6. 5 The QueryRewriting algorithm

Query expansion In the above description of the QueryRewriting algorithm, both the source query Q and the intermediate target query Q are expanded using the ontology constraints, respectively in Lines 2 and 9. This query expansion process, as described by the QueryExpand function of Figure 6.6, uses the strategy of applying the ontology constraints to "chase" the query, similarly to the *chase* algorithm that is used in relational databases to compute dependency implications or optimize queries [1]. In relational databases, a database constraint can be represented as a *tgd* (*tuple generating dependency*) in the form $\forall x \exists y\ \varphi(x) \rightarrow \psi(x, y)$, where φ and ψ are conjunctions of atoms. In an ontology setting, we consider three kinds of constraints, namely, *subclass, subproperty,* and *typing* constraints, all of which can be represented as a *tgd.* Specifically, the *tgd* $\forall x\ C_1(x) \rightarrow C_2(x)$ corresponds to a subclass constraint $C_1 \subseteq C_2$; the *tgd* $\forall x \forall y\ P_1(x, y) \rightarrow P_2(x, y)$ corresponds to a subproperty constraint $P_1 \subseteq P_2$; and the *tgd* $\forall x \forall y\ P(x, y) \rightarrow A(x)$ (resp. $\forall x \forall y\ P(x, y) \rightarrow B(y)$) corresponds to a typing constraint that the instances of *x* (resp.*y*) are of type *A* (resp. *B*).

Similarly to the chase algorithm, QueryExpand is a non-deterministic process that terminates, provided that the dependencies are *acyclic* (we assume no constraints such as $A \subseteq B$, $B \subseteq C$, and $C \subseteq A$ in an ontology) and the applications of dependencies do not introduce new variables into the query (since all the three constraints: *subclass, subproperty,* and *typing* do not contain the existence quantifier). Under these conditions, given a conjunctive query *Q* and constraints Σ over an ontology *O,* it has also been proved that the algorithm QueryExpand has the resulting query Q'= QueryExpand (Q, Σ) equivalent to *Q,* denoted $Q \equiv Q'$ *[1].* This means that the answers to both queries are the same over all the ontology instances that satisfy the constraints. As an example, let us take the preceding query on multiple family land parcels, and denote it by *Q.*

Algorithm QueryExpand (Q, Σ)
Input: a conjunctive query *Q* over ontology *O;* the constraints Σ over *O.*
Output: The query *Q* after the expansion.

1 **Repeat**
2   **Let** ϕ be $body_Q$;
3   **Let** $\psi : R_1(x) \rightarrow R_2(x)$ be any dependency in Σ;
4   **If** there exists a homomorphism *h* from $R_1(x)$ to ϕ, but not from $R_1(x) \wedge R_2(x)$ to ϕ, **then**
5   Extend *h* to a new homomorphism h' from $R_1(x) \wedge R_2(x)$ to ϕ;
6   Add $h'(R_2(x))$ into $body_Q$;
7   **Else** exit repeat;
8 **End repeat**

Fig. 6. 6. The QueryExpand algorithm

As specified on the global ontology *G,* all the properties (e.g., xyCoordinates) referred in *Q* belong to the class LandParcel, thus leading to the corresponding typing constraints. Such constraints can be represented by a *tgd* of the form $\forall x \forall y\ P(x, y) \rightarrow A(x)$ (e.g., $\forall x \forall y\ xyCoordinates(x, y) \rightarrow LandParcel(x)$). By applying them to *Q,* we obtain the following expansion of *Q:*

ans(*a*, *b*, *c*) :– *xyCoordinates*(*x*, *a*), *bounding*(*x*, *b*), *jurisName*(*x*, *c*), *state*(*x*, "Wisconsin"), *luCode*(*x*, "115"), *LandParcel*(*x*).

Furthermore, given that the LandParcel class is a subclass of Land in *G*, the corresponding *tgd* (e.g., $\forall x\ LandParcel(x) \rightarrow Land(x)$) of such constraint is still applicable to the above query. The final resulting expansion Q^* of *Q* is as follows:

ans(*a*, *b*, *c*) :– *xyCoordinates*(*x*, *a*), *bounding*(*x*, *b*), *jurisName*(*x*, *c*), *state*(*x*, "Wisconsin"), *luCode*(*x*, "115"), *LandParcel*(*x*), *Land*(*x*).

Schema-level mapping The key to query rewriting lies in Lines 4 to 7 of the QueryRewriting algorithm, which maps the expanded source query Q^* to a new query Q' over the target ontology, based on the set of schema-level mappings in *M*. Similarly to the ontology constraints used by QueryExpand, ontology mappings can be treated as constraints specified over the source and the target ontologies. Therefore, we express ontology mappings in a *tgd*. However, the use of these mappings is different from the use of ontology constraints for query expansion, as explained in what follows.

Consider two ontologies O_1 and O_2. Given the *tgd* $\psi : \forall x\ R_1(x) \rightarrow R_2(x)$, if ψ represents an ontology constraint constraint $R_1 \subseteq R_2$, where $R_1, R_2 \in O_1$, and $R_1(x)$ is part of the query's body, then ψ is applicable to the query, and $R_2(x)$ should be added to the query. This query expansion, as described by the QueryExpand function, will not bring false positives to the query's answer, since the instances that satisfy ψ are in O_1. In comparison, if ψ is an ontology mapping $R_1 \subseteq R_2$, where $R_1 \in O_1$ and $R_2 \in O_2$, then this constraint implies a potential data transfer from O_1 to O_2. In this sense, $\psi: \forall x\ R_1(x) \rightarrow R_2(x)$ is not applicable to queries containing R_1 (like in the ontology constraint case), but is applicable to those containing R_2. This happens because a query retrieving instances of R_2 is also retrieving instances of R_1, given the semantics of ψ.

Therefore, the application of a dependency $\psi: R_2(x) \rightarrow R_1(x)$ to a query *Q*, as Line 7 of QueryRewriting indicates, is performed by taking the converse ψ' of ψ (i.e., $R_1(x) \rightarrow R_2(x)$), followed by the operations specified in Lines 4 and 5 of QueryExpand. The resulting $R(x)$ (in Line 8 of QueryRewriting) is then $h'(R_2(x))$ as in Line 6 of QueryExpand. The following shows the result of mapping Q^* (the expanded source query) to a query Q' on the local ontology O_1 according to the mapping *M* as presented in Figure 6.4:

ans(*a*, *b*, *c*) :– *xyCoordinates*(*x*, *a*), *boundingBox*(*x*, *b*), *jurisName*(*x*, *c*), *state*(*x*, "Wisconsin"), *lu*1 (*x*, "115"), *LandParcel*(*x*).

If we compare Q' to the previous two queries (Q and Q^*) obtained in the query rewriting process, we notice that *Land*(*x*) was first added into Q' by the query expansion step, and then it disappeared after the schema-level query mapping. In reality, the *LandParcel*(*x*) in Q^* is different from *LandParcel*(*x*) in Q':the former is against the global ontology G, whereas the latter is against the local ontology O_1, as shown in Figure 6.4. Therefore, the disappearance of *LandParcel(x)* from Q' is due to the mapping from *LandParcel(x)* and *Land(x)* on G to *LandParcel(x)* on O_1.

Instance-level mapping Both the QueryExpand function and the query mapping process are performed at the schema level. In comparison, the rewriting of the constants that are referred to in the query is based on the instance-level mappings between two ontologies, particularly the mappings between two land use taxonomies. We describe next the instance rewriting process of Figure 6.7.

In this case, we have c = {"Wisconsin", "115"}. From the mapping between G and O_1 shown in Figure 6.3, it follows that RT$\subseteq$115 and RM $\subseteq$ 115. Therefore, from Lines 3 to 9, we have that A_1 = {"Wisconsin"} and A_2 = {"RT", "RM"}. Now that we have two vectors of constants (c' in the algorithm): {"Wisconsin", "RT"} and {"Wisconsin", "RM"}, we obtain the union of the following two queries (see Lines 10 to 13).

ans(a, b, c) :– xyCoordinates(x, a), boundingBox(x, b), jurisName(x, c), state(x, "Wisconsin"), *lu*1 *(x,* "RT"), *LandParcel(x).*

ans(a, b, c) :– xyCoordinates(x, a), boundingBox(x, b), jurisName(x, c), state(x, "Wisconsin"), *lu*1 *(x,* "RM"), *LandParcel(x).*

Discussion We have assumed that the schema-level mapping M between two ontologies is a *full mapping,* that is, all relation atoms (including classes and properties) in the body of the query need to have been mapped to some atom in the other ontology, with the mapping type being $\supseteq$ or $\equiv$.Under this assumption, we can prove *the soundness* of the QueryRewriting algorithm on its computation of a rewriting (i.e., target query) ϑ *contained* in the source query Q, denoted $\vartheta \subseteq Q$. A proof sketch follows.

Let Q^* be the expanded source query, Q' be the intermediate target query, Q'' be the expanded intermediate target query. Given that $Q \equiv Q^*$, $Q' \equiv Q''$, and $Q'' \equiv$ *Q[1]*, it suffices to prove that $Q' \subseteq Q^*$. Suppose that t is an instance in the answer to Q', *i.e.,* $t \in Q'$(O), where O is the local ontology instance. Then t makes every predicate $R(x)$ in $body_{Q'}$ true. According to Lines 5 to 8 of the QueryRewriting algorithm, every predicate $S(x)$ in $body_{Q^*}$ is also made true by t. This means that $t \in Q^*$ (G), where G is the global ontology**Error! Bookmark not defined.** instance, therefore $Q' \subseteq Q^*$. We note that we obtain a *contained* rewriting, instead of a *maximally contained* rewriting [23].This is actually due to our preference for high precision rather than for high recall, which we discuss below.

Algorithm ConstantMapping (Q, M)
Input: a conjunctive query Q over ontology O' with constants c_1, ..., c_n from O; the instance level mappings M between ontologies O and O'.
Output: a union Q of conjunctive queries Q' with constants from O'.

1 $\vartheta = \varnothing$;
2 $c=(c_1,\ldots,c_n)$;
3 **For** each c_i, with $i \in [1..n]$
4 $\quad A_i=\{\}$;
5 $\quad$ **Let** C be the class standing for c_i;
6 $\quad$ **For** each $C \supseteq C'$ or $C=C'$ in M
7 $\quad\quad A_i = A_i \cup \{c'\}$, where c' is the constant represented by C';
8 $\quad$ **If** there is no $C \supseteq C'$ or $C = C'$ in M then
9 $\quad\quad A_i =\{c\}$;
10 **For** each $c' \in A_1 \times \ldots \times A_n$
11 $\quad Q' = Q$;
12 $\quad$ Substitute c in Q' with c';
13 $\quad \vartheta = \vartheta \cup Q'$;

Fig. 6. 7. The ConstantMapping algorithm.

There are two important steps involved in the local-to-local query rewriting: *query conversion* and *mapping composition*. The query conversion deals with the conversion of a query (e.g., in XPath) native to the local system to a query (in c-RQL) on the local ontology. However, c-RQL can only represent a particular class of XML queries that have the same expressive power as c-RQL. Therefore we only consider such XPath queries. The other step has to do with the transitivity of the mappings. That is, the composition of two equivalent mappings yields an equivalent mapping. The composition of two subclass mappings (or one subclass and one equivalent mappings) results in a subclass mapping. In the same way, the composition of two superclass mappings (or one superclass and one equivalent mappings) results in a superclass mapping. We do not consider any other mapping compositions.

The last issue we discuss relates to the trade-off between the precision and recall of the query processing. Currently, the the query rewriting algorithm only uses mappings that guarantee the correctness of the query. For instance, given a query Q: $\{x|A(x)\}$, our query rewriting algorithm only rewrites Q to $\{x|B(x)\}$ in two cases: $A \supseteq B$ or $A \equiv B$. This ensures that we will not return to the user instances that do not belong to A. But we may miss some instances of B that are also instances of A and should be included in the answer to Q, thus lowering recall. An alternative is to allow the approximate semantic relationship and to assign a score between [0..1] to every mapping based on the similarity of the mapped classes or properties. Thus, query rewriting can calculate an estimated precision of the target query. In practice, different scenarios impose different requirements on the mappings. For example, an eCommerce application involving purchase orders requires a very precise and complete translation of a query, whereas a search engine usually does not require an exact transformation [8].

6 User Interfaces

In this section, we briefly describe the two user interfaces that assist respectively in ontology alignment (and in particular instance-level mapping) and in query processing.

6.1 Visual Ontology Alignment

The AgreementMaker is a visual software tool that is used to create the mappings between the global ontology and each local ontology and to generate the agreement documents [9]. With this tool, users load two ontologies side-by-side and display each one as a tree of concepts as shown in Figure 6.8. The global ontology is displayed on the left hand side and the local (target) ontology is displayed on the right hand side. Concept (or class) names are displayed in rectangular nodes on the ontological trees.

The AgreementMaker implements a *hybrid* ontology matching process [28], by combining several matching methods to determine the schema-level and instance-level mappings between both ontologies.

The first two criteria are the *rule-based deduction* and the *user interaction,* as discussed in Section 4. User interventions are necessary when deductions are not applicable. Users map concepts manually based on their knowledge of the domain represented by the ontologies. The particular mappings that are established depend on the perceived semantic relationships among concepts. The deduction process automates the creation of new mappings based on existing mappings, provided that the preconditions for a particular deduction are satisfied. The degree of automation depends on the graph topologies and on the degree of similarity between the ontologies [10]. In the case where the topologies differ substantially and deduction cannot be used, the burden on users to perform mappings manually increases [25].

In addition to the previous criteria, the tool also provides *matching by definition,* which matches the name and the description of the concepts. The procedure consults a dictionary (e.g., WordNet[36]) and returns a semantic relationship (e.g., *hypernym, hyponym,* or *synonym)* between both concepts and a *similarity* score ranging from 0 to 100. The shortcoming of this matching criterion is that two concepts can have the same name and the same description, but they could be semantically mismatched because they oc-

[36] http://wordnet.princeton.edu/

cur in different contexts. To address this problem our tool considers the paths leading to the concepts [10].

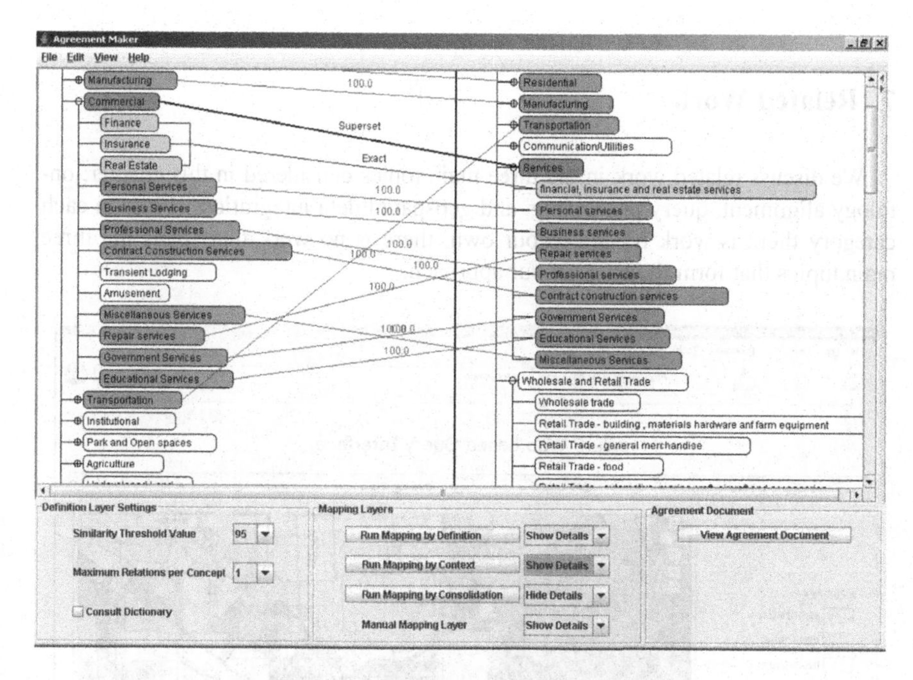

Fig. 6. 8. The ontology alignment interface.

The tool supports a fourth matching step, *matching by consolidation,* whereby users provide a ranking of the matching criteria. In this way, wherever there are conflicting results for the matchings, the highest ranked criterion will take precedence.

6.2 Web-based Query Interface

The prototype of a Web-based visual query interface has been implemented for browsing different types of land usage in a geospatial area that can span several local data sources. This interface serves as a proof-of-concept for the interoperability of heterogeneous geospatial data based on the query rewriting algorithm discussed in Section 5.

Figure 6.9 shows the land use map of the city of Madison where parcels are highlighted with colors indicating their associated land use categories. In particular, the user initiates a query by selecting a State, County, and City, which is then represented by a c-RQL query on the global ontology. The global query is rewritten into subqueries over the local data sources. The results are integrated and visualized in the form of a land use map, which is superimposed on satellite images obtained from Google Map.[37]

7 Related Work

We discuss related work in the three main topics considered in this chapter: ontology alignment, query processing, and geospatial data integration. While in each category there is work related to our own, there is no work that spans the three main topics that form the basis of our approach.

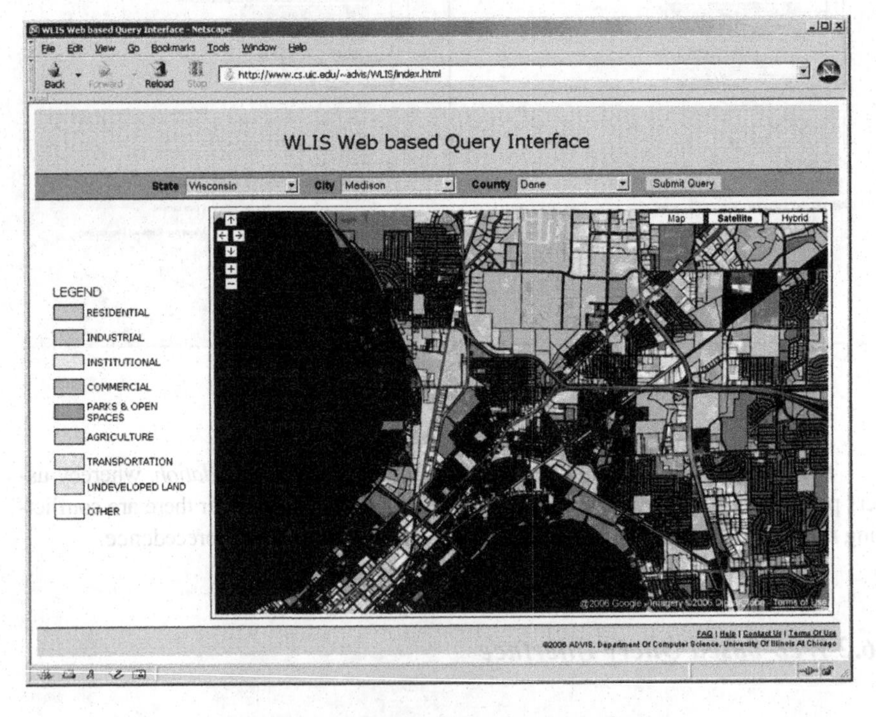

Fig. 6. 9. A Web-based user query interface.

Ontology alignment The problem of ontology alignment has received much attention recently [30]. A large effort has been devoted to techniques that enable the automatic (or semi-automatic) alignment of concepts across ontologies [15]. Existing

[37] http://maps.google.com

alignment approaches make use of one or more ontology alignment (or matching) techniques belonging to the following three categories:

Element-level At the element level, matching can use various similarity measures based, for example, on names of elements or their textual descriptions. A normalized numerical value is calculated for each of the matching candidates, and the best one is selected [4, 6, 26].

Structure-level The structure-level information that can be used by the matching process include the *graph* or *taxonomy* underlying the schema or ontology. Graphs are used as contextual information to map pairs of elements and the taxonomy can provide the matching process with more semantics [24, 25]. An example of semantic-level matching determines the similarity of two concepts based on the similarities of their ancestors [29]. In our approach, we consider the semantic similarity of the concepts' children, instead.

Instance-level Instance-level matching uses the actual contents (or instances) of the schema or ontology elements [12,21].

Although we have two types of ontologies to align, namely the *schema-like* ontologies and the *taxonomy-like* ontologies, in this chapter we have mainly concentrated on the alignment of the latter type. Taxonomy-like ontologies use only the subconcept (or subclass) relationships between two entities, therefore they lend themselves well to the use of structure-level methods. However, element-level and instance-level approaches can be used in conjunction with our structure-level methods. In fact, our prototype makes use of element-level alignment as described in Section 6.

Related systems to ours include Clio[20], COMA++ [3], and Falcon-AO [22]. The first two place a strong emphasis on the user interface, like we do, while the last two share with our approach their support for structure-level automatic matching methods.

Query processing When mappings are defined as (relational) views, query processing is often referred in the literature as view-based query answering or rewriting [19]. However, few view-based query processing algorithms address the issue of query rewriting over ontologies involving the specific kinds of issues involved, which we must take into account [31].

Schema and ontology-based query processing techniques have been proposed both for centralized [2, 27, 34] and for peer-to-peer architectures [5, 13, 16]. While most of these approaches focus on XML or relational query languages to perform the query rewriting, we use RQL because of our choice of metadata and data representation.

Our ontology-based query rewriting algorithm is similar to the *compute WTA* algorithm proposed by Calvanese *et al.* for query reformulation [5] as both assume consistent ontology mappings. However, we allow for the transformation of the values that are contained in the query based on the instance-level ontology mappings. In this way, we can address semantic heterogeneity, which occurs in the land use codes.

Another related approach considers constraint-based query processing in the Clio system [34]. It focuses on schema mapping and data transformation between nested (XML) schemas and relational databases by taking advantage of the schema semantics to generate consistent translations from source to target and by considering the constraints and structure of the target schema.

Geospatial data integration Information integration methodology from the database community has been applied to spatial information systems. For example, the MIX framework offers a meditation approach for integrating heterogeneous sources containing spatial data types (e.g., vector graphics, maps) and associated data (e.g., text, tables, figures, images) [18], which supports a wide range of spatial applications. The system architecture comprises three layers: a foundation layer consisting of databases and wrappers, a mediation layer supporting query and result exchange among the wrapped sources, and an application/user interface layer. In the foundation layer, the data model is exported from the sources in the form of an XML DTD. In the case of spatial information, for example, the wrapper constructs the DTD by using the associated catalog information. The wrappers support scripts that execute complex queries as a combination of several primitive queries. As compared to our approach, semantic relationships are supported, for example in the form of spatial predicates such as *within (region1,region2),* but there is not an overall "semantic graph" that would support, for example, the alignment of spatial attributes.

VirGIS is a more recent approach for mediation of geographical information systems [14]. It differs from MIX in that it adopts newer standards such as GML (Geography Markup Language) for data modeling and WFS (Web Feature Servers) to perform communications (e.g., queries) with clients. It supports mappings between attributes or classes but no semantic overall framework is presented.

8 Conclusions

In this chapter, we focused on data integration and interoperability across distributed geospatial data sources. To illustrate the impact of our approach we showed practical examples that are derived from land use applications.

We propose an ontology-based approach to achieve the integration and interoperability of the distributed geospatial data sources by solving both schematic and semantic heterogeneities. Two different kinds of mappings are established between the global ontology (which describes the domain) and each local ontology (which describes each data source): *schema mappings* between the schema of both ontologies and *instance mappings* between the (land use) taxonomies of both ontologies.

We have discussed two modes of query processing in our system, *global-tolocal* and *local-to-local* (or *peer-to-peer).* Query rewriting in both modes uses the previously established mappings. We propose a c-RQL (conjunctive RQL) query rewriting algorithm, such that the resulting target query is *contained* in the source query, thus providing sound answers to the source query.

Future work will focus on:

- Ontology alignment, and in particular the deduction-based method. Currently, we make some assumptions on the topology of the ontologies. Without such assumptions, we may need to consider the combination of our bottom-up deduction process with top-down reasoning on mappings (e.g., [29]).
- Query rewriting, so as to take into account "approximate" mappings. In this case, precision and recall of query answering will depend on the similarity of the underlying mappings, thus making the ability to determine mapping similarities a critical task.

Acknowledgments We would like to thank Nancy Wiegand and Steve Ventura, from the Land Information & Computer Graphics Facility at the University of Wisconsin-Madison, and the members of WLIS for discussions on land use and other scenarios related to geospatial data integration. We would also like to thank Sujan Bathala, Nalin Makar, Afsheen Rajendran, and William Sunna for their help with the design and implementation of the user interfaces.

References

1. S. Abiteboul, R. Hull, and V. Vianu. *Foundations of Databases*. Addison-Wesley, 1995.
2. B. Amann, C. Beeri, I. Fundulaki, and M. Scholl. Querying XML Sources Using an Ontology-Based Mediator. In Confederated International Conferences DOA, CoopIS and ODBASE, volume 2519 of Lecture Notes in Computer Science, pages 429–448. Springer, 2002.
3. D. Aumueller, H. H. Do, S. Massmann, and E. Rahm. Schema and Ontology Matching with COMA++. In *ACM SIGMOD International Conference on Management of Data*, pages 906–908, 2005.
4. S. Bergamaschi, S. Castano, and M. Vincini. Semantic Integration of Semistructured and Structured Data Sources. *SIGMOD Record*, 28(1):54–59, 1999.
5. D. Calvanese, G. D. Giacomo, D.Lembo, M. Lenzerini, and R. Rosati. What to Ask to a Peer: Ontology-based Query Reformulation. In *International Conference on Principles of Knowledge Representation and Reasoning (KR)*, pages 469–478, 2004.
6. S. Castano, V. D. Antonellis, and S. D. C. di Vimercati. Global Viewing of Heterogeneous Data Sources. *IEEE Transactions on Knowledge and Data Engineering*, 13(2):277–297, 2001.
7. V. Christophides, G. Karvounarakis, I. Koffina, G. Kokkinidis, A. Magkanaraki, D. Plexousakis, G. Serfiotis, and V. Tannen. The ICS-FORTH SWIM: A Powerful Semantic Web Integration Middleware. In *International Workshop on Semantic Web and Databases (SWDB)*, pages 381–393, 2003.
8. V. Cross. Uncertainty in the Automation of Ontology Matching. In *International Symposium on Uncertainty Modeling and Analysis (ISUMA)*, pages 135–140, 2003.
9. I. F. Cruz, W. Sunna, and A. Chaudhry. Semi-Automatic Ontology Alignmentfor Geospatial Data Integration. In International Conference on Geographic Information Science (GIScience), volume 3234 of Lecture Notes in Computer Science, pages 5 1–66. Springer, 2004.

10. I. F. Cruz, W. G. Sunna, and K. Ayloo. Concept Level Matching of Geospatial Ontologies. In *GIS Planet International Conference and Exhibition on Geographic Information,* 2005.
11. I. F. Cruz, H. Xiao, and F. Hsu. An Ontology-based Framework for Semantic Interoperability between XML Sources. In International Database Applications and Engineering Symposium (IDEAS), pages 217–226, July 2004.
12. A. Doan, J. Madhavan, P. Domingos, and A. Y. Halevy. Learning to Map between Ontologies on the Semantic Web. In *International World Wide Web Conference (WWW),* pages 662–673, 2002.
13. M. Ehrig, C. Tempich, J. Broekstra, F. van Harmelen, M. Sabou, R. Siebes, S. Staab, and H. Stuckenschmidt. SWAP - Ontology-based Knowledge Management with Peer-to-Peer Technology. In *German Workshop on Ontology-based Knowledge Management (WOW),* 2003.
14. M. Essid, F.-M. Colonna, O.Boucelma, and A. Bétari. Querying Mediated Geographic Data Sources. In *International Conference on Extending Database Technology (EDBT),* volume 3896 of *Lecture Notes in Computer Science,* pages 1176–1181. Springer, 2006.
15. J. Euzenat, A. Isaac, C. Meilicke, P. Shvaiko, H. Stuckenschmidt, O. ˇSváb, V. Svátek, W. R. van Hage, and M. Yatskevich. First Results of the Ontology Evaluation Initiative 2007. In *Second ISWC International Workshop on Ontology Matching.* CEUR-WS, 2007.
16. E. Franconi, G. M. Kuper, A. Lopatenko, and I. Zaihrayeu. A Distributed Algorithm for Robust Data Sharing and Updates in P2P Database Networks. In *Current Trends in Database Technology - EDBT 2004 Workshops,* Lecture Notes in Computer Science, pages 446–455. Springer, 2004.
17. T.R. Gruber. A Translation Approach to Portable Ontology Specifications. *Knowledge Acquisition, 5(2):* 199–220, 1993.
18. A. Gupta, R. Marciano, I.Zaslavsky, and C. K. Baru. Integrating GIS and Imagery Through XML-Based Information Mediation. In International Workshop on Integrated Spatial Databases (ISD), Selected Papers, volume 1737 of Lecture Notes in Computer Science, pages 211–234. Springer, 1999.
19. A. Y. Halevy. Answering Queries Using Views: A Survey. *VLDB Journal,* 10(4):270–294, 2001.
20. M. A. Hernández, R. J. Miller, and L. M. Haas. Clio: A Semi-Automatic Tool For Schema Mapping (demo). In *ACM SIGMOD International Conference on Management of Data,* page 607, 2001.
21. R. Ichise, H. Takeda, and S. Honiden. Rule Induction for Concept Hierarchy Alignment. In *IJCAI Workshop on Ontologies and Information Sharing,* 2001.
22. N. Jian, W.Hu, G. Cheng, and Y. Qu. Falcon-AO: Aligning Ontologies with Falcon. In *K- CAP 2005 Workshop on Integrating Ontologies.*CEUR Workshop Proceedings 156, 2005.
23. M. Lenzerini. Data Integration: A Theoretical Perspective. In *ACM SIGMODSIGACT-SIGART Symposium on Principles of Database Systems (PODS),* pages 233–246, 2002.
24. S. Melnik, H. Garcia-Molina, and E. Rahm. Similarity Flooding: A Versatile Graph Matching Algorithm and Its Application to Schema Matching. In *IEEE International Conference on Data Engineering (ICDE),* pages 117–128, 2002.
25. N. F. Noy and M. A. Musen. Anchor-PROMPT: Using Non-local Context for Semantic Matching. In *IJCAI Workshop on Ontologies and Information Sharing,* 2001.
26. L.Palopoli, D. Saccà, and D. Ursino. An Automatic Techniques for DetectingType Conflicts in Database Schemes. In *International Conference on Information and Knowledge Management (CIKM),* pages 306–313, 1998.
27. M. Peim, E. Franconi, N. W. Paton, and C. A. Goble. Query Processing with Description Logic Ontologies Over Object-Wrapped Databases.In *International Conference on Scientific and Statistical Database Management (SSDBM),* pages 27–36, 2002.
28. E. Rahm and P. A. Bernstein. A Survey of Approaches to Automatic Schema Matching. *VLDB Journal,* 10(4):334–350, 2001.
29. M.A. Rodr´ıguez and M. J. Egenhofer. Determining Semantic Similarity among Entity Classes from Different Ontologies. *IEEE Transactions on Knowledge and Data Engineering,* 15(2) :442–456, 2003.

30. P. Shvaiko and J. Euzenat. A Survey of Schema-Based Matching Approaches. In *Journal on Data Semantics IV*, volume 3730 of *Lecture Notes in Computer Science*, pages 146–171. Springer, 2005.
31. H. Stuckenschmidt. Query Processing on the Semantic Web. *Künstliche Intelligenz (KI)*, 17(3):22–26, 2003.
32. H. Wache, T. Vögele, U. Visser, H. Stuckenschmidt, G. Schuster, H. Neumann, and S. Hübner. Ontology-Based Integration of Information - A Survey of Existing Approaches. In *IJCAI Workshop on Ontologies and Information Sharing*, 2001.
33. M.Wiegand, D. Patterson, N. Zhou, S. Ventura, and I.F. Cruz. Querying Heterogeneous Land Use Data: Problems and Potential. In *National Conference for Digital Government Research (dg. o)*, pages 115–121, 2002.
34. C.Yu and L. Popa. Constraint-Based XML Query Rewriting For Data Integration. In *ACM SIGMOD International Conference on Management of Data*, pages 371–382, 2004.

Chapter 7: Translating Vernacular Terms into Geographical Locations

Juan José García Adeva

Department of Languages and Computer Systems
Faculty of Computer Science
University of the Basque Country
Donostia – San Sebastián, Spain

jjga@ehu.es

Abstract. Vernacular terms are often used by people when referring to geographical location names. Being able to determine the canonical geographical form of a location from its various other forms and in particular from its vernacular expression would extend the number of queries that can be successfully answered by Web-based mapping applications. There are two basic scenarios when dealing with this type of vernacular geography: i) when there is a known list of vernacular geographical names to translate, so that it is possible to find their descriptions (e.g. in the Web) and use them to infer the canonical name, and ii) no information about vernacular names exists or is known beforehand until query time. This chapter focuses on the latter situation and introduces an approach based on statistical text-mining techniques coupled with a knowledge set, which is a compilation of information retrieved from the Web describing all the known locations within a geographical area (e.g. city, state, country). Translating a vernacular term consists of finding the partition of the knowledge set where this vernacular term occurs and has the highest relevance. This approach is simpler than other existing methods while offering promising results and other advantages such as being language independent. Its experimental implementation consists of a Web service that can be used by any SOAP-enabled application.

1 Introduction

People use different types of geographical languages to describe locations. When searching for known public places (e.g. 25 Pitt street in Sydney, Australia) they use *formal* or *standard* vocabularies. A different scenario is when people use *private* location names (e.g. the pub), only meaningful within a group of related people. An additional way of referring to a place is using *vernacular* terms, which is the subject of this work. This usually happens when people communicate in relation to a particular region or cultural group. Examples include "downtown" (in the USA) or "CBD" (in New Zealand or Australia) to allude to the centre of a city, and "the coathanger" as a synonym of the Sydney Bridge. Another instance is "the strip" to point out a particular section of Sunset Boulevard in Los Angeles, California.

This is somewhat related to what is called *vague spatial concepts*, encompassing spatial relations or regions [24] (e.g. near, to the south, US Midwest, etc.) where there are neither precise boundaries nor exact criteria for membership. In consequence, people's perspective of the concept rarely translates into the digital definition stored in the mapping system. An industry exists around this problem, where companies like Urban Mapping[38] are selling their user-generated expertise to major mapping engines. Other community-oriented approaches such as Wiki-Mapia[39] benefit from volunteers submitting their own boundary definitions of a concept.

However, this work is primarily focused on the particular case where the geometric footprint [17] of the vernacular term has the same granularity and extent as the canonical location defined in the mapping system. There is some work in this area [6, 5], which relies on already known vernacular terms and computational linguistics techniques such as word sense disambiguation to perform resolution of term ambiguity. For this, an ontology in the geographical domain (e.g. a subset of WordNet[40] or a geospatial ontology such as GML[41] and SPIRIT) is used [19, 20, 10] to extract synonymy and meronymy relationships between terms. This ontology is envisioned as an extension to traditional gazetteers. The new terms are then used for ontology-based query expansion [12], where the new terms are added to the user's query. Unfortunately, these approaches are not enough in the proposed scenario because many (if not most) of the vernacular terms in question are unknown to any geographical ontology.

Despite of the importance and user demand of properly dealing with unseen vernacular names in the geographical context [34], attention is seldom paid to this type of geographical vocabulary due to the difficulty of translating the vernacular terms into their corresponding standard forms. This task is even more difficult to tackle when the vernacular term to translate is not known until query time, as opposed to having a prepared list of possible terms that can be processed offline. In

38 http://urbanmapping.com
39 http://wikimapia.org
40 http://wordnet.princeton.edu
41 http://opengis.net/gml/

fact, popular Web-based mapping systems, such as Yahoo! Maps[42] and Google Maps,[43] do not support this type of geography at the moment.

This work introduces a Web service that provides the standard location of a given vernacular term using a statistical text mining approach. The focus of this method is on both simplicity and language independence.

The structure of this chapter is as follows. Section 2 starts with some background information that mentions some related work. Next, Section 3 describes the approach based on statistical text mining methods. Section 4 covers the technical details about an experimental implementation, which can be utilised by other applications through a Web service, and some sample results. The chapter concludes with some remarks about the work and a few ideas for future work in Section

In the context of determining georeferences in text documents, Woodruff and Plaunt [36] reported that "although benchmarking is a daunting task, evaluation is extremely signiï¬cant. Consequently, future work should include the development of a benchmark". More than a decade after this statement, no such benchmark has materialised. Similarly, in our context of vernacular geography, because a standard set of vernacular terms for benchmarking does not exist, it is impossible to perform an exhaustive evaluation on the accuracy of the results. The absence of concensus in regards to an evaluation benchmark contributes to a somewhat unprincipled system development procedure, where there is no consensus about what knowledge contributes most to the task and hence applying heuristics whose utility is unknown, leading to a potential waste of resources.
5.

2 Background

Because they are profusely used in Geographic Information Systems (GIS) in general and this work in particular, I describe what a geographical gazetteer is. I also mention some related work that exists for determining the area or the location that corresponds to a vernacular geographical term. The approach is usually based on finding a text description of the vernacular term and then analysing it in order to extract the standard locations that it describes.

42 `http://maps.yahoo.com`
43 `http://maps.google.com`

2.1 Geographical Gazetteers

236863	University of California-Berkeley	school	Alameda	37.86972	-122.25778
1654172	Berkeley Tennis Club	locale	Alameda	37.85917	-122.24056
1675999	Berkeley Yacht Club	locale	Alameda	37.865	-122.31111
1656763	Sproul Plaza	park	Alameda	37.86972	-122.25806
1656769	Tolman Hall	building	Alameda	37.87417	-122.26306
1703690	Berkeley Station	locale	Alameda	37.87111	-122.26722

Fig. 7. 1. Some entries extracted from a California gazetteer. The actual gazetteer contains additional information fields such as the population and other coordinate formats.

A gazetteer is a dictionary of geographical references that supplies information about place names [18]. The types of places usually are schools, universities, churches, parks, stations, ports, airports, towns, cities, rivers, mountains, etc. Examples of information about these place names that can be found include the postal code, population, size, province, coordinate, etc.

Gazetteers are produced by governmental institutions as well as companies in the form of plain text files or relational databases. There is a large number of both public and commercial gazetteers. For example:

- The World Gazetteer[44] is a public gazetteer that provides the country, province, population, coordinates, and rank among all towns within the country of a given city.
- The Global Discovery Gazetteer is a commercial gazetteer from Europa technologies.[45] It currently contains 707,000 unique worldwide locations including urban sprawls, mountains, airports, ports, roads, railways, rivers, and lakes.
- The Gazetteer of Australia[46] is government supported and provides information on the location and spelling of 315,500 geographical names across Australia.
- The public gazetteer provided by the Geographic Names Information System (GNIS)[47] is the one used in this work, more precisely, the concise version for California, that lists information about major physical and cultural features throughout this state. There is an additional version with historical features.

Most gazetteers are provided in the form of a text file with a large number of entries, each one of them describing a location. Fig. 7.1 shows a few entries that correspond to an actual gazetteer of California. The first column indicates the postal code, to be followed by the location name, the location type (such as

---

44 http://www.world-gazetteer.com

45 http://www.europa-tech.com

46 http://www.ga.gov.au/map/names/

47 http://geonames.usgs.gov

school, park, hospital, forest, airport, mine, etc.), the county name, the latitude, and the longitude. This gazetteer of California carries some 105,000 locations divided into 63 types for the whole state of California, USA. For example, around 600 of them correspond to the county of Alameda (where Berkeley is), of which nearly 100 are about Berkeley.

Unfortunately, traditional gazetteers are quite simple and suffer from some limitations. One of them is that they do not contain spatial or semantic relationships between locations. Finding information like synonyms, hyponyms, related, or neighbouring locations of a source location would be a very valuable feature for vernacular geography and other related problems but not currently available in regular gazetteers.

2.2 Related Approaches

There is a popular trend based on computational linguistics techniques for toponym resolution. This task consists of determining the correct geographical location of a textual place name given several possible mappings [22]. For example, there are cities called York in countries like the UK, Germany, Australia, Canada, Greenland, and the United States. While humans are very good at determining from context which is the intended referent, this task is not a simple one to be performed automatically [21] and it bears some resemblance to word sense disambiguation with two main steps [22]: i) obtain candidate referents for a toponym and ii) select the most likely candidate.

Some of the techniques employed in toponym resolution are interesting for vernacular geography, in particular those using approaches based on place name extraction [29, 27, 14, 25]. In order to determine the correct context of the toponym, a list of place names are extracted from unstructured texts describing the toponym. This extraction task is also known as geoparsing in this particular context or named entity recognition in general. In turn, named entity recognition and part-of-speech tagging (where the entities are linguistic elements of a lexical taxonomy) are the two main types of information extraction tasks [1]. A typical information extraction application analyses text documents and presents the user with the specific information from them that the user is interested in. Information extraction has some similarities with text mining in regards to its blurry definition. Different authors have used different definitions, sometimes contradictory. Chen [8] considers that text mining "performs various searching functions, linguistic analysis and categorizations", therefore including information retrieval within text mining. Other researchers [23, 3] also consider information retrieval a text mining functionality. However, Hearst, who proposed a definition for text mining for the first time [16], explicitly excludes information retrieval and information extraction as part of text mining or as a particular functionality or task within it. The latter point of view has gained strength in the last years, when the research community in information extraction has grown considerably. It is important to bear these definitions in mind to understand where the work described fits.

There is some work [35, 28, 2, 26] that employs location names extraction in order to resolve vernacular terms. When a vernacular name has to be mapped into one or more corresponding standard location names, the vernacular term is used to search and retrieve some text (e.g. Web pages) that describes it. One or more well-known standard location names contained in this description are extracted, and then used to delimit the geographical area or position of the vernacular term. The level of granularity (e.g. city, street) of these extracted geographical references will constrain the geographical footprint of the eventually determined region [17].

There are three main techniques for extracting standard location names from text:

1. Internal methods where the text is analysed without relying on external resources. This approach can be based on either a deductive or an inductive method. In the first case, a model of the language and its grammar has to be built, which usually results in a very long and costly process both to develop and maintain. In the second method, the entities are automatically extracted after an algorithm was previously trained. Models using Hidden Markov and Maximum Entropy are the most common. Although these models have proven successful, they must be trained and optimised for each language and type of text supported, with extensive adaptation required to incorporate new languages.
2. External methods that rely on an external resources, usually a gazetteer. The advantage of this approach is that there is no need for previously training the system and its response time is better than when using the internal approach mentioned above. Its main disadvantage is its rigidity as only references stored in the gazetteer using the same (or very close) spelling can be extracted from the text. To increase the most probably needed flexibility of matching terms in the gazetteer that do not correspond textually but do semantically (e.g. *aly* for alley, *av* or *avn* for avenue, etc.), the language model that corresponds to each geographical or cultural region has to be encoded. Although feasible, this is an extremely tedious and complex task.
3. Hybrid methods that combine both internal and external resources.

Each one of these two approaches should usually be capable of offering fairly good accuracy when mapping a vernacular term into its the geographical area. However, their implementation, although attainable, would require a significant programming effort for each country and language supported. Another limitation to be considered for these techniques is the availability of high-quality Web resources that precisely describe the vernacular terms.

3 Statistical Text-Mining Approach

The approach presented in this work is based on statistical text mining. It has the noteworthy advantage of being language and grammar independent, as opposed to the computational linguistics approach described before, which depends heavily on the language model. Furthermore, this approach would be cheaper computationally speaking in most circumstances. An additional minor advantage of this approach is that it should find the coordinate of not only vernacular terms, but also many unknown standard location names (those not registered in the system or under a different name). However, its main disadvantage might be that the accuracy obtained depends on the granularity of the locations provided by the gazetteer, because the vernacular terms are eventually mapped into these standard geographical locations.

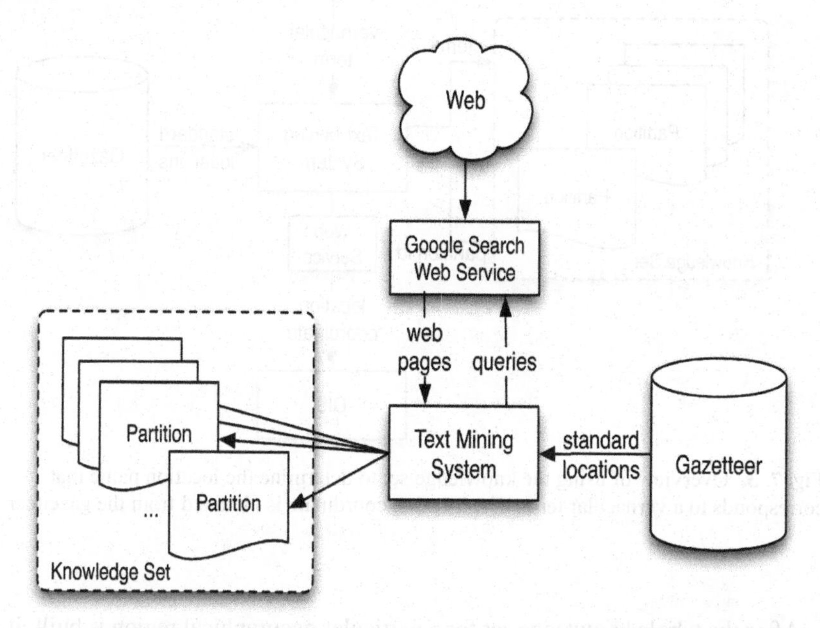

Fig. 7. 2. Overview of the knowledge set creation process, where locations found in a gazetteer are used to query the Web and obtain descriptions.

This technique relies on a *knowledge set* built from external resources. This knowledge set is a compilation of information describing all the locations, within a geographical area (e.g. city, state, country), for which their coordinates are known. Generating this knowledge set is an automatic and one-time process to be completed before the system is ready for use. Building it consists of traversing a list of location names from the gazetteer of the targeted geographical region. Each location name (or entry) in the gazetteer is used to search and retrieve descriptions about that location itself from different sources such as encyclopaedias or the Web. Through the tuning of the search queries plus some additional filtering of the results, only text content that describes the corresponding location will hopefully be retrieved and used to build the knowledge set. All the compiled text documents describing a particular standard location name are combined and used to define the knowledge set partition that corresponds to a particular standard location name. Fig. 7.2 shows a graphical overview of this knowledge set creation process.

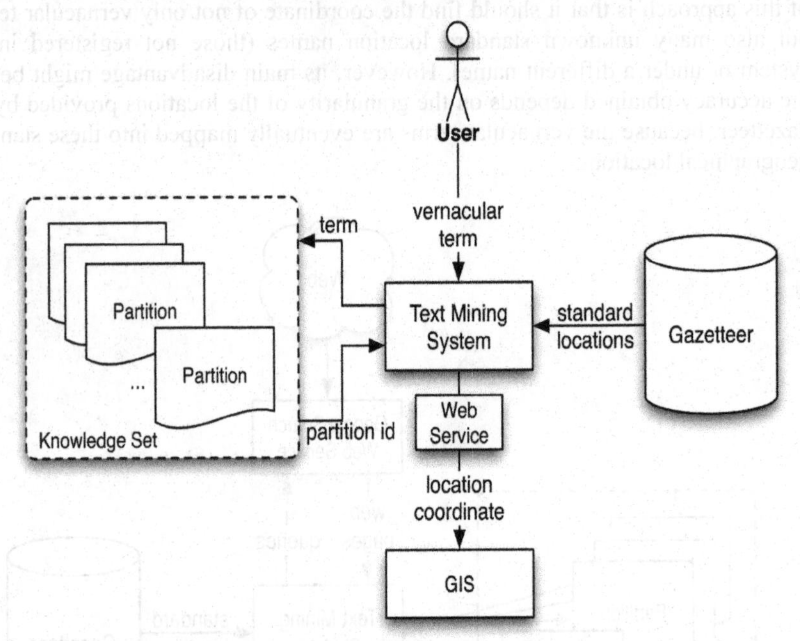

Fig. 7. 3. Overview of using the knowledge set to determine the location name that corresponds to a vernacular term. The location coordinate is obtained from the gazetteer.

After the whole knowledge set for a particular geographical region is built, it is then transformed using the traditional vector space model (VSM) [33]. This is an algebraic model that represents natural language documents and queries in a high-dimensional space, where each dimension of the space corresponds to a word in the document collection. According to this model, for each of the partitions describing an individual location a feature vector in this space is generated.

A large dimension of the feature vectors may affect the effectiveness of the model [7]. Consequently, high dimensionality is mainly avoided through a feature selection process. Feature selection is performed on a dictionary of terms T provided by the preprocessing stage. A new sub-dictionary T' is created where $T' \subset T$

and $|T'|<<|T|$. The reduction factor value is expressed by $\varepsilon = \dfrac{|T|-|T'|}{|T|}$. Each term $t_k \in T$ is weighted using a function that captures the degree of correlation with c_i. A number of terms with the highest weight are used to generate T' and hence to represent the documents from this point on. There are many term evaluation functions studied in the literature [32], many of them from the fields of information theory and statistics. Some of the most common include term frequency, document frequency, inverse document frequency, information gain, mutual information, and χ^2.

For weighting words in this VSM, the tf/idf function (term frequency *vs* inverse document frequency) was selected. The term frequency in the given document offers a measure of the relevance of the term within a document. The document frequency is a measure of the global relevance of the term within a collection of documents.

There are several implementations of the function tf/idf. The one used on this work uses a collection of documents D with length $|D|$, so that $D = \{d_1,\ldots,d_{|D|}\}$. Each document d_j is expressed by a collection of terms with length $|d_j|$, such as $d_j = \{t_1,\ldots,t_{|dj|}\}$. Therefore the tf/idf value for a particular term t_i in a document d_j of D is given by

$$\text{tf/idf}(t_i,d_j,D) = \text{tf}(t_i,d)\cdot \log_2 \text{idf}(t_i,D) =$$

$$\frac{|t_i|}{\max\{\text{tf}(t_1,d_j),\ldots,\text{tf}(t_{|d_j|},d_j)\}}\cdot \log_2 \frac{|D|}{|D\supset t_i|}, \tag{1}$$

where |ti| is the number of times that the term ti occurs in the document dj (which is normalised using the maximum term frequency found in dj), |D| is the number of documents in D, and $|D\supset t_i|$ the number of documents where ti appears.

There are other alternative term weighting approaches to tf/idf such as Robertson/Sparck Jones [30] or OKAPI [31]. Nonetheless, because these methods are usually variations on the way the frequency functions tf and idf are combined, using other term relevance function other than tf/idf should only produce minimal differences on the results.

Once this model has been built, mapping a vernacular term into its most related standard location is a fast and computationally cheap operation. To do so, the feature vector with the highest weight in the dimension that corresponds to the vernacular term is selected, so that the corresponding standard location name and coordinate from the gazetteer can be offered as response. In cases where the vector space model does not have a dimension corresponding to the vernacular term, the mapping is not possible and it can be considered a failure.

Fig. 7.3 offers a graphical representation of this process while a more detailed description can be found in Algorithm 1. It is interesting to note that although this work is only aiming at providing one location as the answer, a number of them could easily be selected instead, hence providing an area as the answer.

Algorithm 1 map(*vernacular_term*)

```
Require: vernacular_term
Ensure: feature_vectors[]
 1: fv ⇐ create_fv(vernacular_term)
 2: max_weights[] ⇐ 0
 3: location ⇐ 0
 4: for all f ∈ feature_vectors[] do
 5:     weights[] ⇐ f − fv
 6:     if weights[] > max_weights[] then
 7:         max_weights[] = weights[]
 8:         location ⇐ f
 9:     end if
10: end for
11: return location
```

4 Experimental implementation

This implementation currently supports the area of Berkeley in California, USA. The gazetteer for California was imported from its tabular text-based format into a relational database based on PostgreSQL.[48]. The database is designed so that each geographical region (e.g. state, country) is stored in a database table.

Building the knowledge set is based on performing Web searches. For this purpose, I chose the Google Search service[49] that provides Web search capabilities to applications through a Web service. Previous registration in order to obtain a free license key is required. Each license key allows up to 1,000 automated queries per day. This limitation is one of the reasons for limiting this implementation to a fairly restricted area. Other similar services include Yahoo! Search[50] and MSN Search. Each gazetteer entry that covers the supported area is used to perform two Web searches. The first search targets all the Web while the rest are limited to articles within online encyclopedias like the popular Wikipedia[51] or Wikitravel.[52] Moreover, the search is performed using either referential phrases or searches

48 http://www.postgresql.org

49 http://code.google.com/apis/ajaxsearch/web.html

50 http://developer.yahoo.com/search/

51 http://wikipedia.org

52 http://wikitravel.org

likely to return pages which are meaningful. A referential phrase attempts to capture geographic relationships. For example, i) containment or inclusion can be represented by a referential phrase such as "*A is a town in B*" where B is the vernacular location name and A is a possible candidate standard location, ii) equivalence can be detected by referential phrases like "*A [also known as | also called | nicknamed] B*".

A number of the top search results (typically 20) are retrieved, filtered to reduce the noise (e.g. removal of listing sections), converted into plain text, and stored in the document repository offered by Pimiento, which will be described in Section 4.2

For example, the following queries would be issued to search for Web pages describing the University of California in Berkeley:

```
"University of California" Berkeley ~California

"University of California * also known as" Berkeley ~California

"University of California" Berkeley ~California site:en.wikipedia.org

"University of California * also known as" Berkeley ~California
site:en.wikipedia.org

"University of California" Berkeley ~California site:wikitravel.org

"University of California * also known as" Berkeley ~California
site:wikitravel.org
```

The operator * means any word in that position and ~ indicates that synonyms of the given word should be included in the results. In this example, issuing `California` is similar to `California OR CA`.

Some of the Web pages obtained with the examples above are not very useful (e.g. the university home page) but the Wikipedia article is very useful. It describes the existence of elements in the campus such as a bell-tower known as Campanile, although its official name is the Sather Tower. Although *campanile* is a term that could appear in other documents not directly related to the University (e.g. Town Hall as indeed happens), the most relevant occurs in the partition of the knowledge set that describes the University of California in Berkeley.

4.1 Text-Mining Software Framework

The implementation of the approach is based on our own text-mining object-oriented application framework called Pimiento [13].

This software component was written using Java Standard Edition (J2SE) and aimed at providing developers with the primary benefits of application frameworks, such as modularity, reusability, extensibility, and inversion of control [11]. The extensibility of the framework is based on a *black-box* approach, where the framework defines interfaces for new components that can be plugged into the framework through object composition. Black-box frameworks are usually structured using object composition and delegation rather than inheritance, as opposed to *white-box* frameworks, which are tightly related to the framework's inheritance hierarchies. In consequence, black-box frameworks are often easier to use and extend than white-box ones. In order to achieve effective reuse, the framework developer must identify those aspects of the target applications that vary from one application to the next, typically called *hot-spots*, and allow explicitly for their variations to be instantiated in applications. In some cases, the hot-spot variations are known by the framework developer, so concrete classes can be provided. Application development then becomes a simple matter of selecting the appropriate concrete classes for the application.

The framework offers the following functionalities in several languages (English, German, French, Spanish, and Basque):

- The **text categorization** functionality includes several learning algorithms such as Naïve Bayes (multinomial and complement), Rocchio, optimal Rocchio, and *k*-Nearest Neighbour. These algorithms can also be used as the base binary learners for ensembles using the decomposition methods one-to-all, pairwise coupling, and error-correcting output codes. Apart from the regular languages supported, a language-neutral preprocessor, based on *n*-grams, is provided for other languages. It includes novel methods for multilingual text categorization, where documents in multiple languages can be found within the same corpus sharing categories. There is also complete evaluation of results including the category-specific measures TP_i, FP_i, FN_i, π_i, ρ_i, F_1 and the averaged measures π^μ, ρ^μ, F_1^μ, π^M, ρ^M, F_1^M , as well as partitioning of the testing space using *n*-fold cross-validation.
- The **language identification** functionality is based on computing and comparing language profiles using character *n*-gram frequencies. A *n*-gram is a chunk of contiguous characters from a word. The number of *n*-grams that can be obtained from a word of length w is w+1. These *n*-grams are used to create a dictionary with the 400 most frequent words ranked by frequency. Therefore, for each supported language, a profile $p_i \in P = \{p_1,\ldots,p_{|P|}\}$ is generated. These language profiles are compared to the profile of a document whose language has to be identified, choosing the language that corresponds to the closest. Comparing two language profiles consists of calculating , where G_i is the set of *n*-

grams in p_i and g_{i_k} the *n*-gram of p_i ranked in *k*-th position. The function $K(g,p_i)$ returns the rank of *g* in p_i. If $g \notin p_i$ then a very high value is used so that $K(g,p_i) >> |G_i|$.

- **Similarity analysis** consists of applying several similarity functions based on distance measures such as Hamming, Euclidean ($\sqrt{\sum_{i=0}^{|S|}(w(s_i)-w(y_i))^2}$), Manhattan ($\sum_{i=0}^{|S|}|w(x_i)-w(y_i)|$), and Minkowski ($\left(\sum_{i=0}^{|S|}|x_i-y_i|^{\lambda}\right)^{\frac{1}{\lambda}}$) to assess the degree of resemblance between two text documents. The documents to compare are previously converted into feature vectors using one of the available weighting methods, including term existence, term frequency, or their combination term frequency/inverse document frequency (tf/idf).
- The **clustering** functionality currently on includes the *k*-means algorithm, which is well known in the literature. This algorithm defines clusters of documents by their corresponding centroid (i.e. centre of mass of the members). A number of iterations are performed where documents are assigned to their closest cluster based on the distance to its centroid. After this, the centroid are calculated again to include the new documents. This algorithms has the weakness often creating singleton clusters for outliers. We applied the method described in [15] to avoid these situations. The distances are measured reusing the similarity functions of the similarity analysis functionality, described above.
- The **summarization** functionality uses a sentence extraction approach, influenced by [9], that uses the following three statistical measures to determine the relevance of sentences: i) because summaries with long sentences are better than those with short sentences, we define *L* as the measure of the sum of sentence lengths (in number of words), where $L(T)=\sum_{i=0}^{n}|s_i|$, ii) because summaries containing sentences found at the beginning of a paragraph are better summaries, the measure *P* is defined by $P(T)=\sum_{i=0}^{n}\frac{1}{1+n}$, and iii) because sentences with high word overlapping probably offer the same information, hence entering redundancy into the summary, the measure *O* is based on the individual term frequencies combined with their inverse sentence frequency

  . In

  these measures, *S* is the text to summarise, with $S = \{s_0, s_1, \ldots, s_n\}$ being the set of sentences that compose it, $s_i = \{t_0, t_1, \ldots, t_{|si|}\}$ the set of terms in sentence s_i, *n* the number of sentences, and $|s_i|$ the number of terms in sentence s_i. Sentences corresponding to questions are not taken into consideration for the summary. These three measures *L*, *P*, and *O* are combined as a weighted sum $Q(S)=\alpha\cdot L(S)+\beta\cdot P(S)+\gamma\cdot O(S)$ that provides an overall score of summarization quality. While this approach is much simpler than other state of the art, linguistic approaches, it offers good results, while being language independent, and computationally very cheap.

4.2 Knowledge Set Persistence

The storage infrastructure for the knowledge set is based on a repository for unstructured text called Lenteja. Although Lenteja is an independent software component that can be used in other applications and systems, it was developed to as Pimiento's text document repository.

A Lenteja repository can use three types of storage for the unstructured text: 1) a relational database, 2) a document-oriented (i.e. XML) database, and 3) a file system. If the relational database is chosen, any database system with an appropriate JDBC driver can be used, although at this moment we primarily use PostgreSQL. In the case of XML databases, we currently support eXist, an open source native XML database. For this project, the relational database was chosen due to the retrieved text documents being completely unstructured.

The repository is organised by collections that contain an unlimited number of documents and collections. Each document can be stored in two different formats: 1) plain text and 2) the OASIS OpenDocument standard [4]. The latter is useful for keeping as much of the original format as possible (e.g. paragraphs and headings) so that these documents can be queried using the XQuery language or modified with the XUpdate language. Each document of collection can be associated with an unlimited amount of metadata. For example, this can be useful for a text categorization functionality to express the categories that the document belongs to.

Lenteja can currently import and export the following document types: PDF, RTF, HTML, Microsoft Word, OpenOffice 1.x, and OpenOffice 2.x.

Applications using Lenteja can use its object-oriented application interface in case they are written in the same language as Lenteja (i.e. Java). For other scenarios, Lenteja currently offers a Web service. Further interoperability mechanism could be added in the future, such as Sun's RMI.

4.3 Application Integration

Web services can be explained as Web sites whose users are software programs instead of humans. The World Wide Web Consortium's (W3C) Web Services Activity[53] created an extension of XML called SOAP. It is used to specify how these

53 `http://w3.org/2002/ws`

programs involved can send and receive parameters to and from remote applications. The interface of the Web service is described with the Web Services Definition Language (WSDL). This allows clients and services to automatically bind. Web services play a major role in a Service-Oriented Architecture (SOA).

The reason for selecting Web services as the network communication mechanism is because Web services encapsulate complexity while allowing the distribution of load and the improvement of scalability. These are common requirements for integrating text-mining techniques into real-world applications. A Web service has three main properties extracted from the above definition: i) its interface contract is platform-independent, ii) it can be dynamically located and invoked, and iii) is self-contained, so that it maintains its own state.

A notable aspect is that due to Web services being self-describing, the client application does not need to know anything about the service except for the format and content of request and response messages. The definition of the message format travels with the message. No external metadata repositories or code generation tools are required.

Using Web services for integration with client applications also presents some limitations and challenges. Sending large collections of documents can present an important performance and scalability issue in scenarios where the network is not controlled or there is only low bandwidth available. This is because the SOAP transport protocol contributes to a high bandwidth usage and processing time (due to the various levels of XML parsing and validation). Securing a Web service requires additional efforts, not only in providing access control in this inherently stateless protocol, but also in avoiding attacks such as denial of service to the server, forged client requests, and interception of messages. In order to avoid interoperability problems between SOAP implementations, only simple types should be used in parameters. This requires an object oriented API to be transformed into its procedural form.

The particular SOAP implementation selected for this project was Apache Axis,[54] which is an open source library based on either Java or C++. One of the interesting features of this software system is its capability to automatically generate the Web service interface in WSDL format using a Java interface.

The application interface is very simple, with just one method called `map` that requires four arguments: vernacular term, city, state, and country. The response by the Web service is an XML message that contains the latitude, the longitude, and an informative message. In case of an error, both the latitude and the longitude are set to zero and the message contains details about the problem.

---

54 `http://ws.apache.org/axis/`

4.4 Experimental results

Table 1 illustrates some examples of query/response that correspond to the area of Berkeley, California, USA. Each query is a vernacular term with its city, state, and country that were sent to the Web service. The response includes the name of the location that contains, or relates to, the vernacular term along with the latitude and longitude in decimal degrees. In the cases shown in Table 7.1 the results were correct and helpful.

Table 7. 1. Some examples of responses by the Web service when queried for vernacular terms within Berkeley, California, USA.

Query	**Response**		
	Latitude	Longitude	Location Name
Shoreline	37.85528	-122.33778	Pier
Downtown	37.87111	-122.26722	Berkeley Station
Amphitheater	37.88528	-122.26167	Rose Garden
Breakwater	37.86806	-122.31528	Berkeley Marina
Cupola	37.86944	-122.2725	City Hall
Spire	37.86944	-122.2725	City Hall

In the context of determining georeferences in text documents, Woodruff and Plaunt [36] reported that "although benchmarking is a daunting task, evaluation is extremely signiï¬cant. Consequently, future work should include the development of a benchmark". More than a decade after this statement, no such benchmark has materialised. Similarly, in our context of vernacular geography, because a standard set of vernacular terms for benchmarking does not exist, it is impossible to perform an exhaustive evaluation on the accuracy of the results. The absence of concensus in regards to an evaluation benchmark contributes to a somewhat unprincipled system development procedure, where there is no consensus about what knowledge contributes most to the task and hence applying heuristics whose utility is unknown, leading to a potential waste of resources.

5 Concluding Remarks

Geographical mapping systems have become very popular thanks to their ubiquity on the Web. Most of the major Internet companies offer free access to their Web-based mapping applications. However, many people that use these systems are not necessarily familiar with the standard geographical terms and thus try to search for vernacular terms that are unknown to the mapping system.

This work offers an approach based on statistical text mining methods and easily available search engines that makes it possible to augment and enrich the available standard geographical vocabulary in order to map these vernacular terms. An experimental implementation of this approach has been implemented, limited to a small geographical region, which is accessible through a Web service. The results obtained by this tool provide an exact location that contains, or is related to, the vernacular term sought. Because of the lack of standard benchmarks available, it was not possible at this time to provide a formal or more extensive evaluation of the system.

Regarding future work, it would be desirable to look into scalability and performance issues when much larger areas are covered (e.g. whole states or countries). A related problem is with gazetteers and their coverage of only parts of the world (e.g. nations or regions), but those few gazetteers that cover the whole world have a much less detailed coverage than those covering less extensive areas. The ADL community[55] proposes exchanging data between gazetteers to overcome these limitations through the ADL Gazetteer Content Standard and the ADL Gazetteer Service Protocol.

This prototype built performs mapping of vernacular with geometric footprints of the same granularity places defined in the gazetteer. It would be possible to enable mapping of vernacular regions in addition or instead, by identifying not only one but several partition of the knowledge set where this vernacular term occurs with a relevance higher than a certain threshold value. However, there are some subtle details like what threshold value to select, how many locations should form the boundary of the region, and how to estimate whether the vernacular term should map into a single location or region. These decisions should be investigated further.

It might also be beneficial to form a committee in charge of producing a standard benchmark. It could be made up by a set of vernacular terms and their corresponding standard geographical locations or areas. This standard benchmark would be very useful to the research community working on this topic.

Acknowledgments Some of this work was done while the author was at the University of Sydney in Australia. The author thanks the anonymous reviewers for their valuable comments on the manuscript.

55 `http://www.alexandria.ucsb.edu/gazetteer/`

References

1. D. Appelt. An introduction to information extraction. Artificial Intelligence Communications, 12(3):161172, 1999.
2. Avi Arampatzis, Marc J. van Kreveld, Iris Reinbacher, Christopher B. Jones, Subodh Vaid, Paul Clough, Hideo Joho, and Mark Sanderson. Web-based delineation of imprecise regions. Journal of Computers, Environments and Urban Systems, 30(4):436–459, 2006.
3. Michael W. Berry. Survey of Text Mining. Springer-Verlag New York, Inc., Secaucus, NJ, USA, 2004.
4. Michael Brauer, Patrick Durusau, Gary Edwards, David Faure, Tom Magliery, and Daniel Vogelheim. Open document format for office applications (OpenDocument) v1.0. Technical report, OASIS Standard, 2005.
5. Davide Buscaldi, Paolo Rosso, and Emilio Sanchis. Wordnet as a geographical information resource. In Proceedings of Third International Wordnet Conference, 2005.
6. Davide Buscaldi, Paolo Rosso, and Emilio Sanchis. A wordnet-based query expansion method for geographical information. In Cross-Language Evaluation Forum in CLEF, 2005.
7. Soumen Chakrabarti, Byron Dom, Rakesh Agrawal, and Prabhakar Raghavan. Scalable feature selection, classification and signature generation for organizing large text databases into hierarchical topic taxonomies. The VLDB Journal, 7(3):163–178, 1998.
8. Hsinchun Chen. Knowledge Management Systems: A Text Mining Perspective. University of Arizona, Tucson, Arizona, 2001.
9. H. P. Edmundson. New methods in automatic extracting. Journal of the ACM, 16(2):264–285, April 1969.
10. Max J. Egenhofer. Toward the semantic geospatial Web. In Proceedings of the 10th ACM international symposium on Advances in geographic information systems, pages 1–4, New York, NY, USA, 2002. ACM Press.
11. Mohamed Fayad and Douglas C. Schmidt. Object-oriented application frameworks. Communications of the ACM, 40(10):32–38, 1997.
12. Gaihua Fu, Christopher Jones, and Alia Abdelmoty. Ontology-based spatial query expansion in information retrieval. In ODBASE: OTM Confederated International Conferences, volume 3761. Springer Berlin, Heidelberg, November 2005.
13. J. J. García Adeva and R. C. Mining Text with Pimiento. IEEE Internet Computing, 10(4):27 – 35, 2006.
14. Claire Grover, Harry Halpin, Ewan Klein, Jochen L. Leidner, Stephen Potter, Sebastian Riedel, Sally Scrutchin, and Richard Tobin. A framework for text mining services. In Simon J. Cox, editor, Proceedings of the UK e-Science Programme All Hands Meeting 2004, pages 878–885, Nottingham, UK, 2004. 31st August-3rd September.
15. Ville Hautamäki, Svetlana Cherednichenko, Ismo Kärkkäinen, Tomi Kinnunen, and Pasi Fränti. Improving k-means by outlier removal. In SCIA, pages 978–987, 2005.
16. Marti A. Hearst. Untangling text data mining. In Proceedings of the 37th conference on Association for Computational Linguistics, pages 3–10, College Park, Maryland, 1999. Association for Computational Linguistics.
17. Linda L. Hill. Core elements of digital gazetteers: Placenames, categories, and footprints. In ECDL '00: Proceedings of the 4th European Conference on Research and Advanced Technology for Digital Libraries, pages 280–290, London, UK, 2000. Springer-Verlag.
18. L.L. Hill, J. Frew, and Q. Zheng. Geographic names - the implementation of a gazetteer in a georeferenced digital library. D-Lib Magazine, 5(1), 1999.

19. Christopher B. Jones, Harith Alani, and Douglas Tudhope. Geographical information retrieval with ontologies of place. In Conference On Spatial Information Theory, COSIT'01, 2001.
20. Christopher B. Jones, R. Purves, A. Ruas, M. Sanderson, M. Sester, M. van Kreveld, and R. Weibel. Spatial information retrieval and geographical ontologies an overview of the spirit project. In SIGIR '02: Proceedings of the 25th annual international ACM SIGIR conference on Research and development in information retrieval, pages 387–388, New York, NY, USA, 2002. ACM.
21. Jochen L. Leidner. Towards a reference corpus for automatic toponym resolution evaluation. In Workshop on Geographic Information Retrieval held at SIGIR-2004, 2004.
22. Jochen L. Leidner. Toponym resolution: A first large-scale comparative evaluation, 2006.
23. Thomas W. Miller. Data and Text Mining: A Business Applications Approach. Prentice Hall, 2004.
24. Daniel R. Montello, Michael F. Goodchild, Jonathon Gottsegen, and Peter Fohl. Where's downtown?: Behavioral methods for determining referents of vague spatial queries. Spatial Cognition and Computation., 2/3(1):185–204, 2003.
25. M. Nissim, C. Matheson, and J. Reid. Recognising geographical entities in Scottish historical documents. In Proceedings of the Workshop on Geographic Information Retrieval at SIGIR 2004, 2004.
26. Robert C. Pasley, Paul D. Clough, and Mark Sanderson. Geo-tagging for imprecise regions of different sizes. In GIR '07: Proceedings of the 4th ACM workshop on Geographical information retrieval, pages 77–82, New York, NY, USA, 2007. ACM.
27. Bruno Pouliquen, Ralf Steinberger, Camelia Ignat, and Tom De Groeve. Geographical information recognition and visualization in texts written in various languages. In Proceedings of the 2004 ACM symposium on Applied computing, pages 1051–1058, New York, NY, USA, 2004. ACM Press.
28. R. Purves, P. Clough, and H. Joho. Identifying imprecise regions for geographic information retrieval using the Web. In Proceedings of the GIS RESEARCH UK 13th Annual Conference, 2005.
29. E. Rauch, M. Bukatin, and K. Baker. A confidence-based framework for disambiguating geographic terms. In A. Kornai and B. Sundheim, editors, HLT-NAACL 2003 Workshop Analysis of Geographic References, Edmontonand Albertaand Canada, 2003. Association for Computational Linguistics.
30. Stephen E. Robertson and Karen Sparck Jones. Relevance weighting of search terms. Document retrieval systems, pages 143–160, 1988.
31. Stephen E. Robertson and S. Walker. Okapi/keenbow at TREC-8. In Proceedings of the eighth Text REtrieval Conference, pages 151–161, 1999.
32. M. Rogati and Y. Yang. High-performing feature selection for text classification. In Proceedings of the eleventh international conference on Information and knowledge management, pages 659–661, New York, NY, USA, 2002. ACM Press.
33. Gerard Salton. Automatic Text Processing: The Transformation, Analysis, and Retrieval of Information by Computer. Addison-Wesley, Reading, Pennsylvania, 1989.
34. M. Sanderson and J. Kohler. Analyzing Geographic Queries. University of Sheffield, 2004.
35. T. Waters and A.J. Evans. Tools for Web-based gis mapping of a fuzzy vernacular geography. In Proceedings of the 7th International Conference on GeoComputation., 2003.
36. Allison Gyle Woodruff and Christian Plaunt. Gipsy: automated geographic indexing of text documents. J. Am. Soc. Inf. Sci., 45(9):645–655, 1994.

Chapter 8: Personalizing Location-Aware Applications

Eoin Mac Aoidh[1], Michela Bertolotto[1], David C. Wilson[2]

[1] School of Computer Science and Informatics, University College Dublin.
{eoin.macaoidh, michela.bertolotto}@ ucd.ie

[2] Department of Software and Information Systems, University of North Carolina at Charlotte.
davils@uncc.edu

Abstract. The knowledge of a client's interests and preferences are invaluable information from a business perspective. Knowledge about a specific client allows the business to tailor their product or service to best suit the client's needs, leading to a healthy relationship between both parties. In addition, knowledge of potential client's interests allow businesses to maximize returns and reduce costs by directing their advertising at the correct target audience. In the spatial domain, knowledge of a client's interests and preferences can be obtained through information on the user's map browsing habits. Our system offers users a map browsing interface and monitors their interactions as they browse the map. Experiments show that the analysis of individual user's interactions reveals implicit interests in specific map areas and particular features. Further analysis leads to the identification of trends and gives rise to inferences as to the user's context. Contextual information about the user allows us to construct a user interest model and gives us the opportunity to apply personalization techniques, targeting the advertising of suitable products and services at potential clients with specific interests. In this chapter we survey personalization techniques in the field of GIS and present a flexible framework for location-aware business applications to implicitly infer user's interests with spatial data from their mouse interactions. Our methodology and some experimental results are also described in detail.

1 Introduction

Detailed information about the target audience, and indeed the ability to distinguish a member of the target audience from an uninterested member of the public is essential for an effective business marketing strategy. An effective audience-aware marketing strategy will achieve a greater return on its investment. This provides our motivation to glean as much information on users' contexts as possible while they browse a spatial dataset. Businesses providing tools handling geographic data have had little opportunity to glean this kind of marketing information in an implicit manner to date. In contrast to vendors in other domains, such as Amazon.com or Google.com, both of whom monitor user interactions to infer their implicit interests, recommending CDs and books in the case of Amazon, and providing adverts for relevant products in the case of Google. In the location-aware business domain, such information on the target audience has had to be obtained explicitly, by surveying system users with questionnaires. Potential clients can find questionnaires irritating and time consuming, which leads to poor completion rates and insufficient, hurried answers. We employ implicit information collection methods (which have a 100% completion rate) to unobtrusively profile system users.

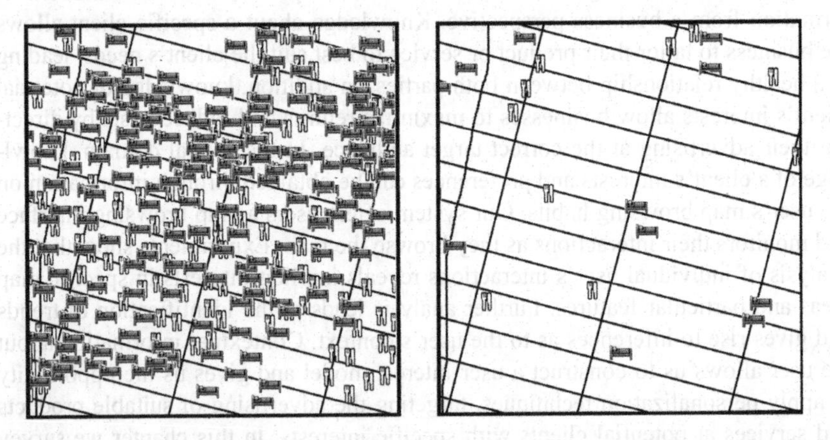

Fig. 8. 1. An impression of a section of New York City, showing lodging and dining facilities. The image on the left shows too many hotels, and too many restaurants; an example of information overload. The image on the right shows how a personalised map might look; showing only budget hotels, and reasonably priced eateries, which are more suited to the user's profile, reducing information overload by eliminating irrelevant expensive hotels and restaurants.

In addition to using profiles of system users for marketing purposes, the information has an important secondary utility, it can also be applied to reduce information overload, making the user's experience with the system as intuitive and as productive as possible. It also paves the way for returning users building an affinity with the system. Information overload happens as a result of the availability of too much data. Figure 8.1 provides an example of spatial information overload. Viewing a map with an application such as Google Earth [1], with the available "lodging and dining facilities" information displayed in a region such as New York provides far more information than any one user could possibly take in. Without employing some method to filter out less relevant data, or a means of ranking the data, the user must sift through large swathes of data in the hope of finding the information he requires.

Information concerning user interests and preferences is collected by the system through implicit methods. It is then used to generate a user interest model. This user interest model is updated on a session by session basis. Thus a profile is maintained for each user which reflects his interests as they change over time. Personalization methods are employed to customise the system to suit the needs of individuals. Products and services suited to the user can be advertised on the interface, increasing his exposure to relevant products and services. Furthermore, the spatial information composing the map can be tailored to the individual, cutting out extraneous detail and even recommending relevant detail; reducing information overload and improving the user's overall experience. Our approach to collecting implicit information about user's interests involves the logging of user's actions as they interact with the system. In particular we focus on mouse movements and clicks. The movement and click information is processed by an interest determining algorithm which we have developed. The algorithm computes an ordered list of every object in the map, based on its relevance to the user for a given session. By maintaining an average of the lists from each session for a particular user, we produce an up-to-date interest model that reflects his changing interests over time.

This chapter surveys user modeling and personalization techniques in the field of GIS in section 2. Section 3 outlines our flexible framework approach for location-aware business applications to implicitly infer user's spatial interests from their mouse movements and describes a series of experiments and results validating our methodology. Section 4 briefly describes the system's implementation. Section 5 addresses personalization to suit the target market. Section 6 discusses the extent of our approach. Plans for future work are presented in section 7.

2 Literature Review

In this section we review research literature items that address some of the issues of concern to our work. The three main tenets of our work are implicit interest indicators, user modeling and personalization. Most of the techniques used in these three areas remain underdeveloped in the spatial domain, thus much of the literature covered in this section pertains to non-spatial Web data. We review this work and comment on its applications in the spatial domain. Spatial data is more complicated than standard pictures and text on a Web page because it presents another problem dimension. The location, shape, scale and relationship among geographical objects are as important as the objects themselves in the perception of spatial data. User modeling and personalization inherently require the collection of information about users, thus we also briefly consider literature documenting the privacy ethics of user modeling.

2.1 Implicit interest indicators

There are two categories of interest indicators; implicit and explicit. Explicit interest indication requires the client to state his interests to the vendor through an explicit means such as completing a questionnaire or applying ratings to items shown to him. This can be irritating and time consuming as it means the client must deviate from his original task. As a result, many clients may avoid such tasks, leaving the vendor clueless as to his interests. Implicit interest indicators on the other hand avoid this problem. The client is uninterrupted. Information is collected based on his actions, and his interests are inferred from these actions.

Claypool et al. [2] discuss implicit interest indicators in detail. Techniques range from keyword extraction from documents, to event logging, such as book marking Web pages, mouse movements, and key-strokes. These events all give a significant indication of interest. For instance, if a client is to bookmark a Web page there is a reasonably high chance that the Web page is of interest to the user. Similarly, if the user clicks on a number of Web links all pertaining to the same subject area, then there is a high chance that the subject area is of interest. Implicit techniques have the advantage of always gathering information without the client having to sacrifice any of his time, or deviate from his task at hand, while explicit techniques are entirely at the mercy of the client. If the client decides not to participate, the vendor gets no information on the client's interests. The shortcoming of implicit techniques is that some indicators are weaker than others, and may be less reliable than a client explicitly stating an interest in something, thus indicator combinations are more desirable than single indicators.

Our approach to gathering implicit interests is founded on a combination of mouse movements, mouse clicks and map browsing actions. A significant amount of work has been carried out in measuring mouse movements as indicators of interest, [3, 4, 2, 5] are examples of such work. The Cheese system [3] records the position of the mouse in relation to Web pages displayed on the screen. Observation of users showed that when pages have lists of information (in text format) 30% of the time, the mouse is employed as a pointer to read along the items in the list. The authors' study also found that some users rest their mouse in white space for fear of accidentally clicking on a link. As an experiment, volunteers were asked to purchase a CD of their choice on Amazon.com using the Cheese system. By analysing their mouse movements as they completed their task, it was possible to infer their second choice of CD to an accuracy of 75%.

Curious Browser's experiments [2] have shown that mouse tracking helped to distinguish a user's amount of interest in a page at a basic level. A more precise degree of interest, and the ability to rank pages in the order of interest was found to be impossible to calculate. Significantly, the authors recorded only the overall amount of time spent moving the mouse, and the number of mouse clicks. The location of the mouse and related clicks was not recorded.

MouseTrack [5] is a Web logging system that tracks mouse movements on Websites. It records all movements and clicks. It is a visualisation tool for Web developers to get a visual depiction of how users interact with their site. To this extent there are no statistical results published to date, however the graphical results show strong evidence to suggest that a user's mouse movements are largely in the area of the page that is of interest to him. Movements and clicks cluster around menu bars and the most popular links on a page. Furthermore, it is possible to distinguish between the arbitrary movements around the page, and the smooth horizontal movements associated with reading lines of text.

Results from the aforementioned mouse-tracking systems all show that it is possible to determine a user's interests to varying degrees of success, depending on the depth of analysis, and number of factors taken into account. On a superficial level, simply producing a heat-map visualisation of mouse movements, clicks, and dwell time give a reasonably good indication of the navigational patterns through such Web sites, and the areas of the screen that are interacted with most by the system users. We modify the techniques used by these systems in order to apply them to spatial data. To the best of our knowledge, no work has been documented to date regarding the correlation between a user's mouse movements and interests with spatial data.

Mouse tracking is en-vogue, as it is functions as a poor man's version of eyetracking. Eye tracking hardware is expensive, and difficult to calibrate, making it impractical for use in most situations. Despite its expense, it has been shown to be extremely effective in determining a person's focus of attention. Gaze position and thought processing have a close correlation, as shown by [6, 7]. Indeed the relationship between thought processing and eye movements has been extended to include mouse movements. Experiments carried out by [8, 6] show that eye movements can be correlated with mouse movements. Different correlations were established between different groups of people by [6], based on the manner in which they use the mouse, ranging from an almost perfect mapping between eye and mouse movements, to users who use only their eyes until they have identified their target before moving their mouse at all. Even at this very loosely correlated level of eye and mouse movements, the correlation between thoughts and movements is still evident. Once again, all research in this area has only been carried out with non-spatial data to the best of our knowledge.

We wish to make our techniques readily available for use by any location-aware application, thus we base our work on the link between thought processing and mouse movements as an implicit interest indicator under the assumption that it also holds for spatial data. By basing our work on mouse movements, rather than eye tracking, our framework is more widely applicable. The value of the information gleaned form mouse movements is strengthened by combining it with information acquired from other interactions linked to thought processing such as mouse clicks and map navigational actions.

2.2 User Modeling

Regardless of the approach used to determine a client's interests, the next challenge presented is how to use the acquired information. The approach we adopt is to produce a user interest model. This allows for data pertaining to a particular client, taken from multiple sessions to be merged to produce a detailed, continuously updated model of the individual. It is these interest models that subsequently provide the required information to personalise the application content for each individual client.

The construction of an interest model requires the analysis of the interaction data captured by a system; it must be interpreted to determine the patterns of interest indicated in the data. A number of techniques have been developed for the mining of interaction data, predominantly for Web sites. Srivastava et al. [9] and Cooley et al. [10] discuss three clearly defined phases in discovering the usage patterns from Web data; data preprocessing, pattern discovery, and pattern analysis. Other publications of note, such as Mobasher et al. [11-13], also pertain to Web data. Some of the techniques applied by the authors are not applicable to spatial data, as it has a different structure with respect to the Web data used by such systems. For instance techniques involving Web links, the use of key words, and the use of cookies are not applicable to a location-aware, spatial application. The core concepts of pattern discovery and user model construction for Web data however provide a detailed and established framework to follow and extend. Extensions provided by our framework include provisions for concepts such as map scale, physical distance between objects, and interaction with point, line and polygon objects.

The CoMPASS system [14] develops user models based on implicit interaction with spatial data. The interest indicators used however are feature-based. The maps provided by CoMPASS are vector maps, which mean features can be added to the map as transparent layers which overlap to make up the map. CoMPASS uses the turning on and off of these layers, in conjunction with map browsing behaviour as implicit indicators of interest. Our framework also provides support for vector map features being turned on/off, and in addition uses mouse movements and browsing behaviour as its main source of interest indicators. CoMPASS's user interest models are also maintained and updated over multiple sessions. In addition to maintaining interest models on an individual basis however, we explore the notion of comparing the interest models against each other, and using techniques such as case based reasoning and regression analysis (discussed in section 5) to identify clients with similar interests. This will allow for the establishment of discrete groups of users based on their contexts.

2.3 Personalization

Application personalization involves tailoring the application to suit a particular user or group of users. The development of adaptive, or personalised systems, which dynamically adapt to the user, make the user's goal easier to achieve. Many of the systems previously mentioned in the implicit interest indicators and user modeling sections perform Web-personalization. In addition to these systems, personalization functionality has also been documented by a number of research projects in the spatial domain, including [14-16]. Many of these systems are mobile-based systems.

Location-aware business applications can be personalised both in terms of the dataset offered to the user, and in terms of the interface within which the data is presented. Deep Map [15] and CRUMPET [16] are mobile systems for spatial data. They provide city tours whose content can be personalised for an individual; however they use explicit methods to collect the individual's preferences. The CoMPASS system [14] personalises the spatial dataset returned to the individual through implicit information collection techniques: User interactions in the form of standard GIS map operations, (pan, zoom etc.) are recorded and used to develop user profiles. We build on such an implicit profiling approach by introducing the analysis of mouse movement data (location, duration, direction etc. of movements) to improve the accuracy of implicit user profiling. In addition we also explore the notion of interface personalization by adapting the layout of the interface to suit individual users.

The notion of interface personalization is more widely exploited on mobile systems than desktop systems. Applications on mobile devices have very limited screen space, and as a result must adapt innovative displays. As many mobile applications involve the use of geospatial data, much of the research involves designing map interfaces. Nivala and Sarjakoski [17] detail the contexts which should be considered when designing an adaptive map interface for a mobile device. The contexts include location, system, purpose of use, time, physical surrounds, navigation history and cultural and social orientation. The interface should be adapted to best suit these contexts. DBHabits is a non-spatial system [18] providing an adaptive user interface which discovers the tasks a user performs by observing his behaviour. The tasks are then made available to the user as macro scripts to improve the ease of future interactions. The authors address both individually adaptive interfaces and group adaptive interfaces, to suit particular kinds of users that can be contextually grouped together. Observing user behaviour to identify the tasks being performed is a pertinent technique to our framework. Not only can the appropriate tools be given priority on the interface, but product advertisements can also be assigned to the most effective areas on the interface in response to the user's interface behaviour. Anderson et al. [19] also developed a non-spatial personalised interface for mobile systems. Their system is designed to re-format Web pages to suit the display of a mobile device. User preferences and device characteristics are taken into account to provide the most intuitive display of Web-based information to the user.

2.4 Privacy

While personalization may provide an improved experience to the user and provide valuable information to the vendor about his target audience, it is worth bearing in mind that privacy is a major concern for most computer users. Kobsa et al. published a number of papers addressing privacy ethics in Web-based personalised systems. Wang and Kobsa [20] completed analysis which reveals that "82-95% of people have refused to give personal information to a Web site at one time or another." and "6-40% of users always supply fictitious information to a Web site when asked to register." These statements apply mainly to the collection of explicit preferences. They cast doubt over the integrity of explicitly collected information. It would be difficult, and counter-intuitive to give false information by means of implicit actions. Implicit indicators have a big advantage over explicit techniques in this regard.

According to Kobsa et al, [20, 21] personalised systems are also subject to legal constraints since they collect personal data. A European directive [22] states that "Value-added (e.g. personalised) services based on traffic or location data require the anonomization of such data or the user's consent." This clause clearly requires the user's consent for any personalization based on interaction logs if the user can be identified. A German law [23] specifies that "Usage data must be erased immediately after each session" These laws apply only if the individuals are identifiable. The framework proposed in this chapter does not require individuals to be identifiable. An anonymous, but unique tag that can be used across multiple sessions is all that is required. The linking of this tag to a concretely identifiable individual is not a necessary part of the framework.

3 Approach

We present our approach to ascertaining the user's interests based on his interactions followed by a description of experiments involving a real dataset in this section. The foundation of our assessment of a user's interests is based on his mouse movements. Our bottom line is that an examination of when and where the user rested his mouse will reveal his interests. We provide additional support for the concepts of panning, zooming, re-centering and adding/removing layers of features to/from the map, all of which change both the displayed content and/or the scale of the map. The application framework supports vector maps, which consist of transparent layers of features overlaid on one another. For example a user visiting a city for a weekend could add the layers 'Hotels' and 'Museums' to the map, while removing the layers 'Houses for sale' and 'Grocery stores'. The notion supported here is that adding a layer to the map indicates an interest in its content, and removing a layer indicates a disinterest. All of these actions constitute indications of implicit interest. We use mouse movements as the primary indicator and strengthen their impact by cross-referencing the other indicators (pan, zoom, re-center, layer removal/addition). The notion of a 'frame' is used to describe the view of the map the user gets. Each pan, zoom and re-center action constitutes the production of a new frame. A session is usually composed of multiple frames.

The following formula provides the means to order all of the objects comprising the map (in a particular frame) by importance to the user in question. Features which the user removed from the map (by removing the layer) are not considered in this formula.

$$(\sum_{0}^{n} ObjScore = \frac{D_{pt}(0)}{Dist(ObjPt(0))} + \frac{D_{pt}(1)}{Dist(ObjPt(1))} ... + \frac{D_{pt}(n)}{Dist(ObjPt(n))}) * F$$

To calculate the score of an object of interest, *Obj*, in a specific frame of a specific session, the distance between *Obj* and each mouse dwelling point, *pt*, within the frame is calculated as *Dist(Obj Pt)*. The dwelling time of the mouse at *pt*, D_{pt}, is divided by *Dist(Obj Pt)*. Each of these fractions are added together for a particular object to give it a score based on all of the mouse movements within the frame. The score for *Obj* is then weighted by the score of the frame, *F*. The score of the frame is a function of the time spent viewing the frame, its scale, and the number of features present in the frame. A frame with 30 features, viewed at a small scale is less likely to contain a discernable specific interest for the user than a frame where the user zoomed in considerably to create a frame with only 5 features. The frame with 5 features would achieve a greater frame score in this case.

By dividing the dwell time of the mouse at *Pt(i)* by *Dist(Obj Pt)*, a measure of *Pt(i)*'s importance in the context of the session is obtained. Using this method, dwelling points furthest from *Obj* will receive the lowest scores. Additionally, a longer dwell time will give a greater value for the numerator, $D_{pt}(i)$, allocating higher scores to points where the mouse rested for a longer duration.

This basic formula is computed for each object in the frame. A session (usually containing multiple frames) requires the exercise to be repeated for each frame. As some objects appear in more than one frame, their scores are added for each frame they appear in, giving a final object score for each object in the map. An object's score is indicative of the amount of interaction associated with it during a session. The level of interaction generally reflects the user's level of interest in the object, thus the objects are ranked according to this score. The highest ranking objects are those of most interest to the user. These inferred interests constitute the information used to create a user interest model.

3.1 Experimental Evaluation

The validity of our approach has been assessed by conducting a series of experiments with a real dataset. The dataset in question consists of 80 point data objects (point landmarks) covering an area of approximately 8km^2. The points correspond to ancient burial tombs in the region of Tarquinia in Italy. A screenshot of the interface provided for the experiments with this dataset is shown in Figure 8.2. Experiment volunteers were required to carry out a set of pre-defined tasks, such as "find Tomb X, which dates to the 10th Century B.C. and compare the objects found in it to the objects found in any tomb from 20B.C." Such information about the tombs and their contents is available by clicking on a tomb. Users wrote the answers to their tasks on an answer sheet. Each of the pre-defined tasks was completed by at least three different users to allow for analysis of the same task as completed by various users. Users' interaction styles vary from person to person. This technique allowed us to find a more average interaction technique. In total, we collected data for 74 complete tasks. Our analysis consisted of comparing the ordered list of objects of interest for a user produced by our algorithm to the tombs named on their answer sheet.

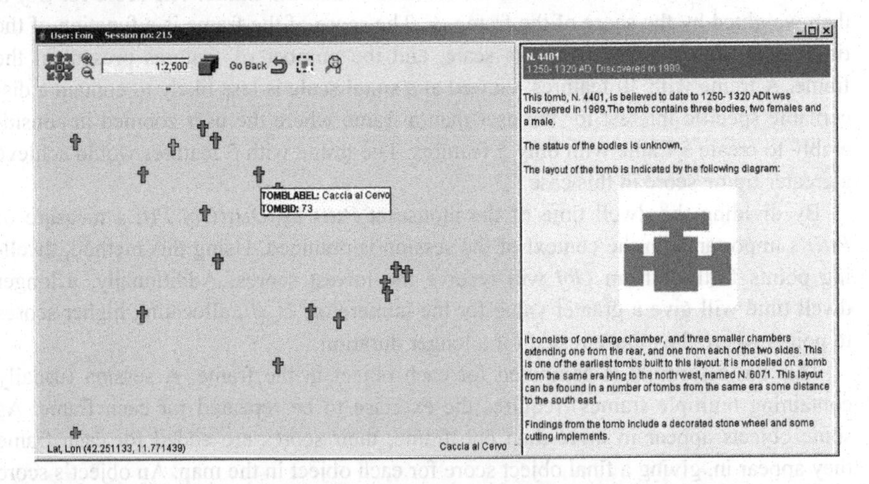

Fig. 8. 2. User interface for experiments: Dataset of burial tombs in Tarquinia, Italy. The user clicks on a tomb on the map (left) to view the associated information (right).

Before presenting the average results of our experiments, we present three metrics developed for evaluation purposes. In order to judge the performance of the algorithm, its accuracy is considered in three areas; Rank Accuracy (RA), Relative Preference (RP) and Absolute Preference (AP).

Rank Accuracy is defined as the accuracy of the rank assigned to the objects in a user's answer. For instance, if a user specifies three objects on his answer sheet, we would expect the algorithm to rank the same three objects given in the user's answer as #1, #2 and #3. If only two of his objects are returned in the top three, then the RA is 66.66%. RA is not concerned with the order of the objects, merely the percentage of them present in the top ranking objects. Essentially RA indicates how accurately the algorithm inferred the user's interests. Relative Preference, RP, is a metric to judge a user's preference for one object in the ranked list over his preference for the following object, based on the object score as computed by the algorithm. It is a measure of the degree of change in preference. For example if object X has a score of 60 and the next highest ranked object, object Y, has a score of 40, RP is computed as (20/60)*100. The degree of change is 33.3%. In other words we can say that the user had a Relative Preference of 33.3% for object X over object Y. The closer the object scores are together, the smaller the RP, thus the less confident we can be that the user had a preference for one object over another. RP is a key metric for distinguishing objects which were of interest to the user from objects which were not of interest. A sudden increase in the RP value represents the divide between objects of interest from those not of interest.

Absolute Preference, AP, is a means of determining the level of interaction associated with any object in an ordered list of interests. It constitutes a mapping of all object scores in the list to a scale of 0 to 100. The object at 0 is ranked lowest, and received the least amount of attention, while the object at 100 is ranked highest, and received the most attention from the user. While RP tells us the degree of preference for one object over another, AP gives an insight into why one object was considered to be preferable to another. It gives an indication as to the quantity of interaction (clicks and/or movements) that was focused around a particular object in the context of all the interactions for all the objects in a particular session.

Table 8.1 shows the average results over all 74 tasks recorded during the experiments. Results for movements alone are shown on the first line; the second line shows the results for a combination of clicks and movements. By combining the indicators, the results are strengthened. Significantly, over all 74 tasks, the system was able to correctly infer the users' interests in 72% of cases based on mouse movements alone. By including information from mouse clicks, the accuracy of determining users' interests rose to 94%.

Table 8. 1. Average experiment results

	RA	RP	AP
mouse movements only	72.30 %	48.97 %	55.89
mouse movements and clicks	94.26 %	85.60 %	57.65

The RP figure taken from each task to calculate the average is the RP of the lowest ranked object given on the users answer sheet. In other words, it reflects the percentage change between the lowest scoring object of interest to the user and all other objects in the map. RP is an important metric to distinguish the fall off in interest, indicating items which are not of interest to the user. By including mouse clicks as an indicator the RP was accentuated from 49% to 86% The AP figure taken from each task to calculate the average is the AP of the lowest ranked object given on the users answer sheet. It indicates the amount of interaction that was focused around that particular object, in the context of all the interactions in the session. It is a metric more useful for examination in individual circumstances than as an average figure.

The average results obtained in this evaluation are encouraging. The results show that mouse movement analysis over spatial data is an accurate technique for inferring users' interests with this particular dataset. (We discuss the characteristics of this technique in relation to other spatial datasets in section 5) It is also apparent from these results that combining implicit interest indicators (i.e. combination of movements and clicks) strengthens the accuracy of inferences made.

4 System Implementation

We describe the configuration of the system used to implement the framework in our experimental scenario in this section. The system is configured to be independent of the dataset used. A full description of the system is available in [24]. The user interface and its functionality are based on OpenMap [25]; an open-source Java based mapping toolkit provided by BBN Technologies. We modify this toolkit to suit our application. The interface produced for the dataset used in the experiments described in the preceding section is shown in Figure 8.2. Interactions are recorded by a transparent interactions layer on the interface.

The spatial datasets are stored remotely in an Oracle 9i Spatial database and are loaded to the interface at the beginning of a session. Upon session termination, the interaction information recorded during the session is transmitted to the database. All data is transmitted and received between components via standard database connections using JDBC, a Java Database Connectivity API for SQL-based operations. Oracle 9i Spatial allows for the execution of complex spatial queries to determine the features that were visible on the user's screen at any point during his session. This information is required for the execution of the interest determining algorithm.

The algorithm produces an ordered list of the user's interests for the session. This list is maintained in the database, and is updated after each subsequent session completed by the same user. By updating the list of interests after each session, the lists keep up to date with the user's interests as they change over time. These up-to-date lists are the user interest models. The user interest models contain the information needed to personalise the application.

5 Target Market Personalization

In this section, we discuss the application of user interest models for location aware personalization to suit the target market. Initially we take the vendors point of view, and consider the application of the user interest model to improve marketing. Then we assess the merits of personalization from a client's point of view, offering a reduction in information overload, and the advertisement of relevant products.

The framework supports two forms of personalization; dataset and interface. Both forms of personalization rely on the interest models generated by monitoring users as they interact with the system. The interest models reflect the best way to filter the dataset. Items in the model with a low score (and items with similar characteristics) can be removed from a personalised dataset, while items with similar characteristics to those items with high scores can be recommended to the user and included in the dataset. Interface personalization can include anything from making certain features appear more prominent (colour and size) to personalizing the arrangement of the tools available on the interface to advertising products on the interface.

From a vendor's point of view, profiling the clients of his system allows him to develop an idea of his target audience. The contents of any advertisements he wishes to display can be compared to the interest models of the user, to find the products and services best matching the client's needs. The placement of the advertisements can be decided based on the areas of the interface that receive the most attention, which may vary from user to user, depending on the arrangement of the interface. Priority can be given to certain aspects of the dataset which the vendor may wish to promote in tandem with his advertisements. For instance if the vendor wishes to advertise 'Hotel Yorba', in addition to placing an advertisement on the interface, he could highlight the 'Hotel Yorba' at the same time on the map by increasing the size of the icon and changing the colour slightly to distinguish it from the other hotels in the dataset.

Clients enjoy an improved interaction experience, as information overload is reduced. This is increasingly important with larger datasets. By way of example, Figure 8.1 shows a screen-shot of the Google Earth application [1] showing all of the 'lodging and dining' facilities available in a small area of New York City. The number of facilities shown could be greatly reduced if it was known that the user was a budget tourist looking for budget accommodation and had no interest in expensive hotels. Unfortunately, some users may find personalization irritating. Especially in instances where they must interact with the system in a way drastically different to their usual interaction characteristics. For instance, if a parent, accustomed to expensive hotels was to research budget hostels for his teenage daughter, this would cause a sudden dramatic change in the information collected for his profile. It is necessary as a result, to provide the option to opt out of user profiling and personalization to the client. Indeed many of the privacy laws also require that this option be made available.

5.1 Group Contexts

The interest models produced by the framework outlined in this chapter are created on an individual basis. They can be used as building blocks to create group models. There are many further extensions possible to the basic user models produced here. By comparing models against each other, patterns could be discovered in clients' contexts. An example of such a pattern might be that clients who show an interest in designer clothing retail stores will also show an interest in exclusive nightclub venues. If a vendor can identify certain patterns in map browsing behaviour then he may be able to predict items which will be of interest to the client based on how other clients with similar interests behaved. Two techniques suited to this kind of pattern analysis are Case Based Reasoning (CBR) and regression analysis. Both of these techniques involve finding patterns in the existing models, and extrapolating from the patterns what interests may develop in the future. For these techniques to work, there must be an existing set of models to work from. The larger the set of models, the more accurate pattern analysis can be. Without any models to work from, these techniques cannot be applied. This is known as 'the cold start problem' [26].

The cold start problem is also evident when a new user interacts with the system for the first time. Generating an interface and dataset dynamically from their user profile is a problem, as their profile has not yet been developed. As a solution to this problem we propose the grouping of users into collaborative 'families' of users. Examples of such families are; a tourist family, a surveyor family and a student family, each family with different information needs and defining characteristics. The family model is a representation of the typical settings that suit the average member of the model. When a new user wishes to interact with the system, the interface suited to his particular user family could be initially presented to the user, however this still leaves the problem of assigning the user to a family. It could be done explicitly, requesting users to provide the seed by selecting the family into which they best fit.

Alternatively this task could also be performed implicitly. It is difficult to glean detailed contextual information on an implicit basis, however, we identified mild contextual information based on user's mouse movements during the experiments described in section 3. Cox and Silva [6] identified three distinct categories of mouse-user during eye-tracking experiments conducted with non-spatial data. 1) Mouse On Side (MOS) where users left their mouse to the side of the menu while their eyes located the target, once the target was located the mouse was moved to the target. 2) Mouse Hovering Target (MHT) Where users hovered their mouse over the target while their eyes scanned the remainder of the menu, and 3) Mouse

With Eyes (MWE) which is characterised by the user's mouse closely following the user's eye movements. Though we do not make use of eye-tracking software, our experiments revealed user categories with parallel traits to the groups identified by Cox and Silva. By producing heat maps of users' interactions, we identified two distinct kinds of user. 1) Lazy mouse users, who only move and click the mouse when absolutely necessary. 2) Fast and frequent mouse users, who make exhaustive use of the mouse, moving it constantly, and clicking impatiently on everything in sight.

More often than not, lazy mouse users corresponded to experienced, efficient users who took less time to complete their task. Fast and frequent mouse users identified themselves as "inexperienced with spatial data" on the experiment's questionnaire. They took more time to complete their tasks often returning repeatedly to areas of the map to double check things. These contexts are not very definitive at present, however there are definite trends evident. With further research and experiments specifically designed to examine these trends, we feel that they could be exploited to glean useful contextual information and assign users to various user families based on their interaction.

6 Dataset Dependency

The accuracy and efficiency of our interest determining method is inherently dependent on the dataset. Given a different distribution of data and/or a dataset containing a mix of different data types, the results of our method may be affected positively or negatively. Figure 8.3 shows a number of different data distribution possibilities. Our original experiments for analysis of the interest determining algorithm described in this chapter were conducted with a real dataset. This dataset is roughly represented by the first section of Figure 8.3. The results of the algorithm were very promising with this dataset, which consisted of a relatively even spread of points, covering an area of approximately 8 km^2. A second real dataset has recently been acquired (roughly represented by the second section of Figure 8.3), consisting of points, lines and polygons, covering a much larger area, almost 7,000 km^2. The distribution of the data is also less evenly spread. Much of the point data such as hotels and shops are focused around a central business district, the road network also converges towards the centre of the most populated area and becomes sparse in the hinterland.

By exploring new datasets with varying data distributions we intend to evaluate our method more thoroughly, determine its strengths and weaknesses, in order make it more robust. A problem that we immediately envisage arising with an uneven spread of data using our current method is outlying points (Such as a scenic viewing area, far from any other points of interest in surrounding towns for instance). Such outlying points could be assigned either too high or too low an object score (indicating their importance to the user), due to their separation from the other evenly spread clusters of data. Another problem that could arise is how best to modify our existing method to estimate a user's interest in line and polygon objects such as roads and parks. In the case of a road, is it best to pick the centroid of the line, or the point on the line closest to the mouse resting point, or a series of points in the line? Polygons give rise to similar questions.

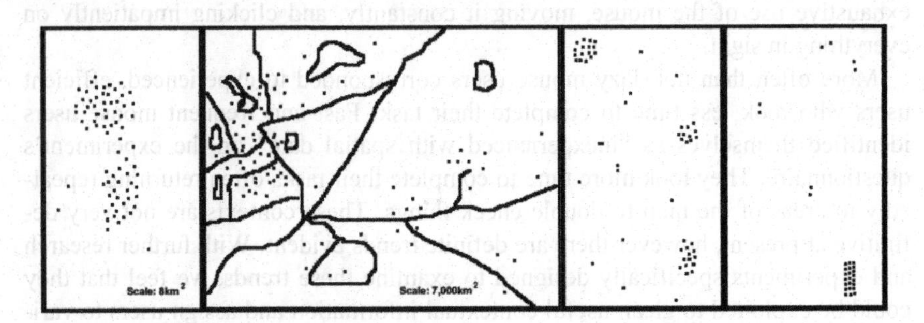

Fig. 8. 3. Dataset distribution examples: (From left to right) An illustration of the distribution of the dataset used in the experiments. An illustration of the dataset containing points lines and polygons for the next experiment (note the uneven distribution). Two example datasets which could be used to test the dependence of the algorithm on the spread of data.

A new set of experiments will ascertain the difference made to algorithm performance by the distribution and type of data used. The final two sections of Figure 8.3 show two example representations of different data distribution scenarios. A number of such datasets will be generated, containing mixes of points, lines and polygons. It is a certainty that the algorithm's performance will be affected by the spread and type of the data, however the spread of spatial data is not a new problem in the field of geospatial services. Spatial indexing is an example of an area of research affected by the same problem. Various different types of indexing systems have been developed for use depending on the spread of the available data. Similarly, We envisage modifications to our algorithm to be necessary given different spreads of data. For example, datasets consisting of a number of clusters such as those illustrated in the final section of Figure 8.3 might be better evaluated as a series of five individual datasets based on the minimum bounding box of each of the clusters before combining the results to give a set of final scores for the importance of each individual point of interest, rather than taking the entire dataset as one. Another modification might be the separation of each dataset into points, lines and polygons, using different rules to assess each individual datatype before combining the results.

7 Future Work

We have demonstrated that the framework developed for interpreting user's interests based on their implicit interactions with spatial data is effective in this instance. The framework is independent of the dataset. Further extensions to the interest determining algorithm are likely to be necessary when datasets incorporating lines and polygons as map features are introduced. An evaluation of how the spread of data affects the performance of the algorithm is also necessary.

While the current research focuses on a desktop framework, initial research on the development of a mobile framework has commenced. Data transfer is more complicated between a mobile device and a remote server, which is necessary to house the Oracle Spatial database. Screen size is also very restricted. Mobile devices do not generally make use of a mouse; they use a stylus instead, however such devices can also include GPS receivers. Stylus use discloses less information than mouse use, as they are not in constant contact with the screen. Their use is analogous to only using the mouse when necessary and removing one's hands from the mouse as soon as it is in the desired location. The introduction of GPS positioning may counter this reduction in access to interaction information. It would be possible to consider the physical location of the user in relation to the data he is viewing as an indicator of interest, under the assumption that he has a greater interest in the things nearer to him than those that are far away. There are many different avenues to explore regarding future work, each of them offering the opportunity to improve the accuracy of interests inferred for users, which becomes increasingly important as the quantity of geographic information becomes more prolific. Better interest interfering techniques will lead to more accurate personalization, improving the experience of both consumer and vendor alike.

References

1. Google earth. http://earth.google.com/, 2007.
2. M. Claypool, P. Le, M. Waseda, and D. Brown. Implicit Interest Indicators. In Proceedings of the International Conference on Intelligent User Interfaces (IUI'01), pages 33-40, Santa Fe, New Mexico, USA, 2001. ACM.
3. F. Mueller and A. Lockerd. Cheese: Tracking Mouse Movement Activity on Websites a Tool for User Modeling. In Proceedings of the Conference on Human Factors in Computing System (CHI'2002), pages 279-280, Seattle, Washington, 2002. ACM.
4. R. Atterer, M. Wnuk, and A. Schmidt. Knowing the Users Every Move - User Activity Tracking for Website Usability Evaluation and Implicit Interaction. In Proceedings of the 15th international conference on World Wide Web, pages 203-212, Edinburgh, Scotland, 2006. ACM.
5. E. Arroyo, T. Selker, and W. Wei. Usability Tool for Analysis of Web Design Using Mouse Tracks. In Conference on Human Factors in Computing Systems CHI'06 Extended Abstracts on Human Factors in Computing Systems, pages 484489, Quebec, Canada, 2006. ACM.

6. A.L. Cox and M.M. Silva. The Role of Mouse Movements in Interactive Search. In Proceedings of the 28th Annual CogSci Conference, pages 1156-1162, Vancouver, Canada, July 26-29 2006.
7. B. Pan, H. Hembrooke, G. Gay, L. Granka, M. Feusner, and J. Newman. The Determinants of Web Page Viewing Behavior: An Eye Tracking Study. In 2004 symposium on Eye tracking research and applications. ETRA, pages 147-154, San Antonio, Texas, 2004.
8. M.C. Chen, J.R Anderson, and M.H Sohn. What Can a Mouse Cursor Tell Us More? Correlation of Eye/Mouse Movements on Web Browsing. In Conference on Human Factors in Computing Systems CHI '01, pages 281-282, Seattle, Washington, 2001. ACM.
9. J. Strivastrava, R.T. Cooley, M. Deshpande, and P-N.Tan. Web Usage Minng: Discovery and Applications of Usage Patterns From Web Data. In SIGKDD Explorations 1(2), pages 12-23. ACM, January 2000.
10. R. Cooley, B. Mobasher, and J. Strivastrava. Data Preparation for Mining World Wide Web Browsing Patterns. In Knowledge and Information Systems 1(1), page 532, 1999.
11. B. Mobasher, H. Dai, T. Luo, and M. Nakagawa. Effective Personalization Based on Association Rule Discovery From Web Usage Data. In Web Information and Data Management, Atlanta, Georgia, USA, November 9-11 2001. ACM.
12. B. Mobasher, H. Dai, T. Luo, Y. Sun, and J. Zhu. Integrating Web Usage and Content Mining for More Effective Personalization. In Proceedings of the First International Conference on Electronic Commerce and Web Technologies, pages 165-176, September 4-6 2000.
13. B. Mobasher, R. Cooley, and J. Srivastava. Automatic Personalization Based on Web Usage Mining. In Communications of the ACM, volume 43, pages 142-151, August 2000.
14. D. Wilson, J. Doyle, J. Weakliam, M. Bertolotto, and D. Lynch. Personalized Maps in Multimodal GIS. International Journal of Web Emerging Technology 3(2):196-216, 2007.
15. R. Malaka and A. Zipf. Deep Map: Challenging IT Research in the Framework of a Tourist Information System. In S. Klein and D. Buhalis, editors, Information and Communication Technologies in Tourism, pages 15-27. Springer, 2000.
16. A. Zipf. User-adaptive Maps for Location Based Services (LBS) for Tourism. In Proceedings of the 9th International Conference for Information and Communication Technologies in Tourism, pages 329-338, Innsbruck, Austria, 2002.
17. A Nivala and L.T. Sarjakoski. Need for Context-Aware Topographic Maps in Mobile Devices. In Proceedings of the 9th Scandinavian Research Conference on Geographic Information Science ScanGIS2003, pages 15-29, Espoo, Finland, June 4-6 2003.

Index